Thermal Systems

Thermal Systems

Editors

Ivan CK Tam
Brian Agnew

MDPI • Basel • Beijing • Wuhan • Barcelona • Belgrade • Manchester • Tokyo • Cluj • Tianjin

Editors
Ivan CK Tam
Associate Professor in Marine Engineering Design & Technology
Newcastle Research & Innovation Institute (NewRIIS)
Newcastle University
Singapore

Brian Agnew
Professor of Energy and the Environment
Newcastle Centre for Railway Research (NewRail)
Newcastle University
UK

Editorial Office
MDPI
St. Alban-Anlage 66
4052 Basel, Switzerland

This is a reprint of articles from the Special Issue published online in the open access journal *Energies* (ISSN 1996-1073) (available at: https://www.mdpi.com/journal/energies/special_issues/TPS).

For citation purposes, cite each article independently as indicated on the article page online and as indicated below:

LastName, A.A.; LastName, B.B.; LastName, C.C. Article Title. *Journal Name* **Year**, *Volume Number*, Page Range.

ISBN 978-3-03943-841-9 (Hbk)
ISBN 978-3-03943-842-6 (PDF)

© 2021 by the authors. Articles in this book are Open Access and distributed under the Creative Commons Attribution (CC BY) license, which allows users to download, copy and build upon published articles, as long as the author and publisher are properly credited, which ensures maximum dissemination and a wider impact of our publications.
The book as a whole is distributed by MDPI under the terms and conditions of the Creative Commons license CC BY-NC-ND.

Contents

About the Editors vii

Ivan CK Tam and Brian Agnew
Thermal Systems—An Overview
Reprinted from: *Energies* **2021**, *14*, 175, doi:10.3390/en14010175 1

Guillermo Valencia Ochoa, Carlos Acevedo Peñaloza and Jorge Duarte Forero
Thermo-Economic Assessment of a Gas Microturbine-Absorption Chiller Trigeneration System under Different Compressor Inlet Air Temperatures
Reprinted from: *Energies* **2019**, *12*, 4643, doi:10.3390/en12244643 5

Guillermo Valencia Ochoa, Jhan Piero Rojas and Jorge Duarte Forero
Advance Exergo-Economic Analysis of a Waste Heat Recovery System Using ORC for a Bottoming Natural Gas Engine
Reprinted from: *Energies* **2020**, *13*, 267, doi:10.3390/en13010267 23

Chuanwei Zhang, Zhan Xia, Bin Wang, Huaibin Gao, Shangrui Chen, Shouchao Zong and Kunxin Luo
A Li-Ion Battery Thermal Management System Combining a Heat Pipe and Thermoelectric Cooler
Reprinted from: *Energies* **2020**, *13*, 841, doi:10.3390/en13040841 41

Ning Qian, Yucan Fu, Marco Marengo, Jiuhua Xu, Jiajia Chen and Fan Jiang
Heat Transport Capacity of an Axial-Rotating Single-Loop Oscillating Heat Pipe for Abrasive-Milling Tools
Reprinted from: *Energies* **2020**, *13*, 2145, doi:10.3390/en13092145 57

Kevin Sartor and Rémi Dickes
Experimental Validation of Heat Transport Modelling in Large Solar Thermal Plants
Reprinted from: *Energies* **2020**, *13*, 2343, doi:10.3390/en13092343 73

Charalampos Alexopoulos, Osama Aljolani, Florian Heberle, Tryfon C. Roumpedakis, Dieter Brüggemann and Sotirios Karellas
Design Evaluation for a Finned-Tube CO_2 Gas Cooler in Residential Applications
Reprinted from: *Energies* **2020**, *13*, 2428, doi:10.3390/en13102428 85

Roberto Barrella, Irene Priego, José Ignacio Linares, Eva Arenas, José Carlos Romero and Efraim Centeno

Reprinted from: *Energies* **2020**, *13*, 2723, doi:10.3390/en13112723 103

Dae Yun Kim, You Na Lee, Joon Han Kim, Yonghee Kim and Young Soo Yoon
Applicability of Swaging as an Alternative for the Fabrication of Accident-Tolerant Fuel Cladding
Reprinted from: *Energies* **2020**, *13*, 3182, doi:10.3390/en13123182 127

Mirosław Grabowski, Sylwia Hożejowska, Beata Maciejewska, Krzysztof Płaczkowski and Mieczysław E. Poniewski
Application of the 2-D Trefftz Method for Identification of Flow Boiling Heat Transfer Coefficient in a Rectangular MiniChannel
Reprinted from: *Energies* **2020**, *13*, 3973, doi:10.3390/en13153973 143

Younghyeon Kim, Seokyeon Im and Jaeyoung Han
A Study on the Application Possibility of the Vehicle Air Conditioning System Using Vortex Tube
Reprinted from: *Energies* **2020**, *13*, 5227, doi:10.3390/en13195227 . . . . **157**

About the Editors

Ivan CK Tam (Associate Professor, Dr.) is an Associate Professor at the Newcastle University with a strong track record of leading innovative and design projects. He has research interest in the combustion process, exhaust emission control, energy management and renewable energy technology. His recent research interest is the application of cryogenic technology in the use of liquefied natural gas and the organic Rankine cycles.

Brian Agnew (Prof. Dr.) joined the Department of Mechanical Engineering at Newcastle University in 1984 with research interest in heat transfer, internal combustion engines and thermal systems. Subsequently, he was appointed as Professor of Energy and the Environment at the School of the Built Environment, Northumbria University. He will continue as a Guest Member of Staff at Newcastle University until September 2022.

Editorial

Thermal Systems—An Overview

Ivan CK Tam [1,*] and Brian Agnew [2]

1. NewRIIS-Newcastle Research & Innovation Institute, Newcastle University, 80 Jurong East Street 21, Singapore #05-04, Singapore
2. NewRail-Newcastle Centre for Railway Research, Newcastle University, Newcastle upon Tyne NE1 7RU, UK; brian.agnew@newcastle.ac.uk
* Correspondence: ivan.tam@newcastle.ac.uk

Received: 3 December 2020; Accepted: 30 December 2020; Published: 31 December 2020

We live in interesting times in which life as we know it is being threatened by human-made changes to the atmosphere we live. On the global scale, concern is focused on climate change due to greenhouse gas emissions and atmospheric pollution produced by combustion processes. The increase in global warming, added to the scarcity of fossil fuels, has motivated the development of new technologies to improve the efficiency of existing processes in power plants. To meet the dual challenges presented by these factors, consideration needs to be given to energy efficiency and pollution reduction in transport and energy conversion processes. A possible approach is through development of new ideas and innovative processes to current practices. Among the available options, multi-generation processes such as trigeneration cycles, battery storage systems, solar power plants and heat pumps have been widely studied as they potentially allow for greater efficiency, lower costs, and reduced emissions. On the other hand, some researchers have been working to increase the potential of energy generation processes through heat recovery with steam generators, organic Rankine cycles, and absorption chillers. This Special Issue is a collection of fundamental or applied and numerical or experimental investigations. Many new concepts in thermal systems and energy utilization are explored, discussed, and published as original research papers in the "Thermal Systems".

The first paper, presented by Ochoa et al. [1], offers an extensive thermo-economic analysis of a heat recovery steam generation system integrated with an absorption refrigeration chiller and a gas micro-turbine. The effect of compressor inlet air temperature on thermo-economic performance of trigeneration systems was studied and analyzed in detail based on a validated model. They found some operational conditions where exergy was highly destroyed due to the exergy inefficiencies of the equipment such as combustion chamber, microturbine, compressor, evaporator, heat exchanger and generator which are found to be important as exergo-economic factors.

In another investigation, Ochoa et al. [2] present an analysis of a waste heat recovery system based on the organic Rankine cycle from the exhaust gases of an internal combustion engine. They studied the exergy destroyed values and the rate of fuel exergy, product exergy, and loss exergy. They found exergo-economic analysis was a powerful method to identify the correct allocation of the irreversibility and the real potential for improvement between components.

Zhang et al. [3] perform an experimental investigation to enhance the working performance and temperature control of electric vehicle batteries through a thermal management system with a heat pipe and thermoelectric cooler. Heat pipes with high thermal conductivity were used to accelerate dissipating heat on the surface of the battery with an additional thermoelectric cooler to increase discharge rate. The findings support the results generated from engineering simulation and show that the combined system can effectively reduce the surface temperature of a battery.

Qian et al. [4] propose the application of oscillating heat pipes to reduce thermal damage in an abrasive milling tool. Heat pipes are passive heat transfer devices with excellent heat transport capacity and they are applied to the machining process to enhance heat transfer. The experimental investigation studied the effects of centrifugal acceleration, heat flux, and working fluids, hence, methanol, acetone, and water, on their thermal performance. Based on their theoretical analysis, centrifugal acceleration will increase the resistance for the vapor to penetrate through the liquid slugs to form an annular flow, which was supported by slow-motion visualization. The phase change occurs, and vapor moves to the condenser to release heat by condensing into liquid.

Sartor and Dickes [5] validated numerical results obtained from a heat transport model with experiments in large solar thermal plants at the Plataforma Solar de Almeria in Spain. They argued that the previous work done had limitations in the assessment of temperatures and computational time required for simulating large pipe networks. They proposed to model the dynamic behavior of the whole system based on a few data inputs. Some atmospheric conditions, such as local clouds, could have significant influence on the outlet temperature and other dynamic behavior of the solar field. An alternative method was used to validate a solar thermal plant considering the thermal solar gain and the inertia of the pipes in their investigation. The accuracy of the model was found to be similar to those of the one-dimensional finite volume method with a reduced simulation time.

Alexopoulos et al. [6] validate design procedure from a simulation model with an experimental study of an air finned tube CO_2 gas cooler. Based on the model, the evaluation of various physical parameters such as length and diameter of tubes as well as ambient temperature was conducted. The researchers attempted to identify the most suitable design in terms of pressure losses and required heat exchange area for selected operational conditions. Hence, a simulation model of the gas cooler was developed and validated experimentally by comparing the overall heat transfer coefficient. The comparison between the model and the experimental results showed a satisfactory convergence for selected operational conditions.

Barrella et al. [7] present a feasibility study which analyzed the use of a centralized electrically driven air source heat pump for space heating. Two models were developed to obtain variables in the hourly thermal energy demand and the off-design heat pump performance. The proposed heat pump is driven by a motor with variable rotational speed to modulate the heating capacity in an efficient way. An eco-friendly refrigerant (R290 or propane) was selected for the heat pump. A back-up system was used to meet the peak demand. Renewable energy used via the heat pump showed significant reduction in CO_2 which would otherwise have been produced via normal fossil fuel consumption. The researchers claimed that these results showed that the proposed technology was among the most promising measures for addressing energy demand in vulnerable households.

Kim et al. [8] propose an alternative method of swaging which is claimed to be more efficient than the traditional coating technology in the fabrication of accident-tolerant fuel cladding. In their study, it was found that the specimen exhibited a pseudo-single tube structure with higher thermal stability. They reported that the specimen had a uniform and well-bonded interface structure under optical microscopy and scanning electron microscopy images. The specimen did not show significant structural collapse, even after being stored at 1200 °C for one hour. The experimental results show that tube process has a high potential for development of an ATF cladding with a length of several meters with their geometries calculated according to the design.

Grabowski et al. [9] present a numeric heat transfer investigation validated with an experimental study in flow boiling of water through an asymmetrically heated, rectangular and horizontal mini-channel, with transparent side walls. The mathematical model assumed the heat transfer process in the measurement module to be steady-state with temperature independent thermal properties of solids and flowing fluid. Grabowski et al. applied laminar characteristic flow in the Reynolds numbers study. The experimental data taken were temperatures at strategic points, volume flux of flowing water, inlet pressure and pressure drop, current, and the voltage drop in the heater power supply. They defined two inverse heat transfer problems which were solved by the meshless Trefftz method with two sets of T-functions.

Kim et al. [10] demonstrate the use of vortex tube in an air conditioning system with the objective to get rid of the use of refrigerant gas. The success of the eco-friendly technology will avoid environmental impact due to refrigerant. The vortex tube is a temperature separation system capable of separating air at low and high temperatures with compressed air. In their experimental study, both direct and indirect heat exchange were investigated to test low-temperature air flow rate according to temperature and pressure. The direct heat exchange method was found to have low flow resistance, and ease in control of temperature and flow-rate. As a result, it is judged to be a more feasible method for use in air-conditioning system by the authors.

The papers in this special issue reveal an exciting area, namely the "Thermal Systems" that is continuing to grow. The pursuit of work in this area requires expertise in thermal and fluid dynamics, system design, and numerical analysis as well as experimental validation. We are extremely delighted to be invited as the Guest Editors of this "Special Issue". We have received great support from many colleagues and top researchers of prestigious universities and research institutions. We are heartened to see such a contribution with the aim of tackling the environmental impact or providing low-cost energy options to the humble communities. We firmly believe that, with the continuing collaboration of all researchers, we can enhance our contribution to tackling the numerous challenges faced by global society. We hope that this Special Issue helps to bring the research community into closer contact with each other. Finally, we would like to thank all our authors, reviewers, and editorial staff who have contributed to this publication. I am sure all readers of this Special Issue of Energies will find the scientific manuscripts interesting and beneficial to their research work in the years to come.

Funding: This research received no external funding.

Conflicts of Interest: The authors declare no conflict of interest.

References

1. Ochoa, G.V.; Peñaloza, C.A.; Forero, J.D. Thermo-Economic Assessment of a Gas Microturbine-Absorption Chiller Trigeneration System under Different Compressor Inlet Air Temperatures. *Energies* **2019**, *12*, 4643. [CrossRef]
2. Ochoa, G.V.; Rojas, J.P.; Forero, J.D. Advance Exergo-Economic Analysis of a Waste Heat Recovery System Using ORC for a Bottoming Natural Gas Engine. *Energies* **2020**, *13*, 267. [CrossRef]
3. Zhang, C.; Xia, Z.; Wang, B.; Gao, H.; Chen, S.; Luo, K. A Li-Ion Battery Thermal Management System Combining a Heat Pipe and Thermoelectric Cooler. *Energies* **2020**, *13*, 841. [CrossRef]
4. Qian, N.; Fu, Y.; Marengo, M.; Xu, J.; Chen, J.; Jiang, F. Heat Transport Capacity of an Axial-Rotating Single-Loop Oscillating Heat Pipe for Abrasive-Milling Tools. *Energies* **2020**, *13*, 2145. [CrossRef]
5. Sartor, K.; Dickes, R. Experimental Validation of Heat Transport Modelling in Large Thermal Power Plants. *Energies* **2020**, *13*, 2343. [CrossRef]
6. Alexopolous, C.; Aljonani, O.; Heberle, F.; Roumpedaki, T.C.; Bruggemann, D.; Karellas, S. Design Evaluation for a Finned-tube CO_2 Gas Cooler in Residential Application. *Energies* **2020**, *13*, 2428. [CrossRef]
7. Barrella, R.; Priego, I.; Linares, J.I.; Arenas, E.; Romero, J.C.; Centeno, E. Feasibility Study of a Centralised Electrically Driven Air Source Heat Pump Water Heater to Face Energy Poverty in Block Dwellings in Madrid (Spain). *Energies* **2020**, *13*, 2723. [CrossRef]
8. Kim, D.Y.; Lee, Y.N.; Kim, J.H.; Kim, Y.; Yoon, Y.S. Applicability of Swaging as an Alternative for the Fabrication of Accident-Tolerant Fuel Cladding. *Energies* **2020**, *13*, 3182. [CrossRef]

9. Grabowski, M.; Hozejowska, S.; Maciejewska, B.; Placzkowski, K.; Poniewski, M.E. Application of the 2-D Trefftz Method for Identification of Flow Boiling Heat Transfer Coefficient in a Rectangular Minichannel. *Energies* **2020**, *13*, 3973. [CrossRef]
10. Kim, Y.; Im, S.; Han, J. A Study of the Application Possibility of the Vehicle Air Conditioning System Using Vortex Tube. *Energies* **2020**, *13*, 5227. [CrossRef]

© 2020 by the authors. Licensee MDPI, Basel, Switzerland. This article is an open access article distributed under the terms and conditions of the Creative Commons Attribution (CC BY) license (http://creativecommons.org/licenses/by/4.0/).

Article

Thermo-Economic Assessment of a Gas Microturbine-Absorption Chiller Trigeneration System under Different Compressor Inlet Air Temperatures

Guillermo Valencia Ochoa [1],*, Carlos Acevedo Peñaloza [2] and Jorge Duarte Forero [1]

[1] Programa de Ingeniería Mecánica, Universidad del Atlántico, Carrera 30 Número 8-49, Puerto Colombia, Barranquilla 080007, Colombia; jorgeduarte@mail.uniatlantico.edu.co
[2] Facultad de Ingeniería, Universidad Francisco de Paula Santander, Avenida Gran Colombia No. 12E-96, Cúcuta 540003, Colombia; carloshumbertoap@ufps.edu.co
* Correspondence: guillermoevalencia@mail.uniatlantico.edu.co; Tel.: +57-5-324-94-31

Received: 5 November 2019; Accepted: 3 December 2019; Published: 6 December 2019

Abstract: This manuscript presents a thermo-economic analysis for a trigeneration system integrated by an absorption refrigeration chiller, a gas microturbine, and the heat recovery steam generation subsystem. The effect of the compressor inlet air temperature on the thermo-economic performance of the trigeneration system was studied and analyzed in detail based on a validated model. Then, we determined the critical operating conditions for which the trigeneration system presents the greatest exergy destruction, producing an increase in the costs associated with loss of exergy, relative costs, and operation and maintenance costs. The results also show that the combustion chamber of the gas microturbine is the component with the greatest exergy destruction (29.24%), followed by the generator of the absorption refrigeration chiller (26.25%). In addition, the compressor inlet air temperature increases from 305.15 K to 315.15 K, causing a decrease in the relative cost difference of the evaporator (21.63%). Likewise, the exergo-economic factor in the heat exchanger and generator presented an increase of 6.53% and 2.84%, respectively.

Keywords: thermo-economic assessment; exergy analysis; trigeneration system; gas microturbine; absorption chiller

1. Introduction

The increase in global warming, adding to the scarcity of fossil fuels, has motivated the development of new technologies to improve the efficiency of existing processes in power plants [1,2]. Among the available options, multi-generation processes such as the trigeneration cycle have been widely used as they allow for greater efficiency, lower costs, and reduced emissions [3]. Therefore, researchers have been working to increase the potential of this type of energy generation process through heat recovery under the steam generator, organic Rankine cycles [3], and absorption chillers [4,5].

Absorption chiller cooling technology is increasingly used because it utilizes refrigerants and absorbents that do not have a negative effect on the environment. In addition, it is possible to feed this type of device with waste heat or some other renewable energy source such as solar energy [6]. Therefore, they are systems widely used in the industrial sector because of the lower energy cost production and potential gas emission reduction [7].

Several studies have developed relevant contributions to the thermo-economic analysis and optimization of absorption refrigeration systems [8], but few are related to the trigeneration system working at the different operating conditions. These studies mainly involve the application of the thermodynamic second law to conduct the evaluation and thermal analysis of the system, which is based on the exergy approach [9]. This method allows us to measure the work potential or quality of different forms of energy with respect to environmental conditions [10,11]. Therefore, the environmental condition plays a key role in the thermo-economic performance of thermal cycles.

Kaynakli and Kilic [12] analyzed the effect of an H_2O-LiBr absorption refrigeration system (ARS) on operating conditions by means of the first and second laws of thermodynamics. It was observed that there is an increase in system performance with the increase in generator temperatures and a decrease in condenser and absorber temperatures. However, the effect of the integration of the ARS with the exhaust gases of a thermal prime mover was not studied, and the value of the generator temperature should be determined. In contrast, Martinez and Rivera [13] conducted an energy and exergy analysis for a dual absorption system using the H_2O-LiBr as a working fluid and also concluded that higher generator and evaporator temperatures and lower absorber temperatures lead to improved system performance. Then, a change in inlet air temperature means different operation conditions on the ARS, and a substantial effect on the thermo-economic indicator of the trigeneration systems, such as the relative cost difference and exergo-economic factor.

Consequently, Kaushik and Arora [14] developed an energetic and exergetic analysis of the single and double effect of a cooling absorption system with parallel free water flow. According to the presented results, the coefficient of performance (COP) presented for the single effect ARS was ranging from 0.6 to 0.75, while in the case of the double effect the COP increased from 1 to 1.28, as a result of different operating temperatures of the heat source and evaporator.

In order to identify the exergetic improvement potential in the H_2O-LiBr double effect, ARS, Gomri and Hakimi [15] conducted an energetic and exergetic analysis, calculating the exergy destruction of the system components. They concluded that the absorber and the high-pressure generator are the components that most influence the total exergy destruction of the system. On the other hand, renewable energy had been used as a heat source of refrigeration systems to increase global thermal efficiency. Hence, Rosiek [16] studied a cooling system integrated into a flat plate solar collector, and the results demonstrated that it is possible to obtain the best results from the exergetic viewpoint supplying water to the absorption cooler in a temperature range of 70–80 °C.

A novel configuration was proposed by Pourfayaz et al. [17], by means of an exergetic analysis to increase the overall performance of the ARS, for a fuel cell cooling system in which nanofluids were used as absorbers.

There are different trigeneration systems, which can be classified mainly according to their driving force, the amount of energy used, and the size of the plant. Each of these classifications has a series of classifications, which have certain advantages and disadvantages regarding the acquisition cost, installation, maintenance, operation ranges, necessary conditions, among others [18].

The availability of sources for electricity generation and global warming are alarming factors that lead to concern about the sustainability of energy production in the future, which brings with it the transcendental impact to design more efficient energy systems [19]. Combined Heat, Cold, and Power (CHCP) are some of the alternative technologies to address problems such as growing energy demand, rising energy costs, the security of energy supply, and large environmental impact [20]. Thus, it is presented as a solution with relevant technical potential, economic, and ecological benefits, which allow reducing the use of primary energy sources to energy generation [21]. The trigeneration system is composed of five main elements: primary engine, electric generator, waste heat recovery system, thermal activation equipment, and a control unit [22].

A promising alternative to trigeneration systems that address the energy problem is based on the use of low-capacity primary sources, also called small-scale technologies, which deliver power between 28 and 200 kW [23], such as the gas microturbine considered in this study. These systems are particularly suitable for applications in commercial buildings, hospitals, schools, local industries, office blocks, and single or multi-family residential buildings [24].

Although some research results based primarily on exergetic analysis show an increase in the COP of the ARS, it is not a complete enough analysis to design a thermal system and ignores the economic part of the system. Therefore, the exergo-economic aspect is necessary to incorporate both exergetic and economic analysis into the system. In this way, it is possible to have a better guide for the thermal study of the components [25,26]. Therefore, the optimization of the ARS performance by means of the thermo-economic assessment was applied [27].

Some trigeneration systems had been studied in industrial and commercial applications. The thermo-economic potential of a trigeneration biomass plant was studied [28], using different configurations, parameters, both economic and operational. The exergetic simulation allowed to determine a 72.8% of the energy efficiency, and the exergetic efficiency ranging from 20.8% to 21.1%, but a parametric case studied is not presented to determine the relevant parameters of the trigeneration process. Also, a complete study was conducted considering some performance energetic, economic and environmental indicators, where the performance of a steam turbine trigeneration system for large buildings based on the energy demands of the facility was calculated, and the results were compared with conventional power generation systems [29]. The results show a decrease in the primary energy saving of 12.1%, CO emission reduction of 2.6%, and CO_2 emission reduction of 2.6%. However, a thermo-economic model was not proposed in this research to identify exergy destruction opportunities.

On the other hand, some thermo-economic studies using a chiller in the trigeneration system have been considered, but the use of a gas microturbine as prime mover operating in a trigeneration system is not reported in the literature. Therefore, the integration of an absorption chiller to a trigeneration system was proposed to generate the required energy, and thus assess energy costs and savings, obtaining an annual cost of $US 384,300 per year, and a payback period of 1.8 years [30]. In addition, an economic analysis of a trigeneration system based on a LiBr chiller was developed. The results were compared with respect to other heat and cold generation systems, and the primary energy consumption decreased by 26.6% with respect to cogeneration.

In the case of the trigeneration system using gas turbines as a prime mover, Ahmadi et al. [31] presented energy and exergetic analysis in a trigeneration system with a combined gas turbine cycle. The results showed a greater exergetic destruction in the combustion chamber, in addition to the environmental impact assessment, where the thermal energy efficiency increase 75.5%, the thermal exergetic efficiency increase 47.5%, and the emission of the CO_2 decrease to 158 kg/MWh.

To increase the performance of the trigeneration system, an exergo-economic optimization was conducted using an evolutionary algorithm, where the economic indicators used to optimize the systems are the total revenue requirement and the total cost of the system [32]. The optimization result of the system allows an improvement of 0.207 $/s in the objective function studied, which is 15% lower than the value in the base case.

From this literary review, the main contribution of this paper is to present a parametric study conducted in a trigeneration system integrated by a Li-Br ARS, a gas microturbine, and a waste heat recovery, to study the effect of inlet gas compressor temperature on the energy, exergy and termo-economic indicator. In addition, the analysis includes the application of the energy, exergy balance, exergy destruction calculation, cost balances application, and the thermo-economic modeling by components, considering in detail the acquisition, maintenance, and operating costs.

2. Methodology

2.1. Description of the System

The physical structure of the trigeneration system is presented in Figure 1. Starting with the gas power cycle, where ambient air at atmospheric pressure (state 1) enters to the compressor, where it is compressed, and its pressure rises, from which it goes out to the preheater (state 2) where it interacts with the exhaust gases of the turbine to increase its temperature and obtain a better combustion.

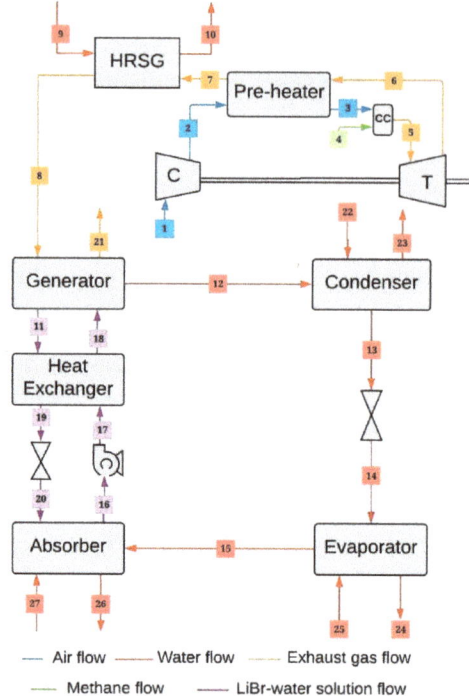

Figure 1. Physical structure of the trigeneration system.

In the combustion chamber, the air flow (state 3) enters from the preheater at high temperature and pressure, and the methane flow (state 4) enters, which will be burned during mixing with excess air. The modeling of the combustion chamber is obtained, assuming an expansion process of the air, which corresponds to an isobaric process inside the system. The resulting combustion gases (state 5) move the turbine in which the hot gases expand and cool rapidly through an adiabatic expansion, generating the power required for the compressor and the net power of the system.

The output gases of the turbine (state 6) are directed to the preheater equipment where its temperature decreases, and then in the Heat Recovery Steam Generator (HRSG) the heat transfer process allows us to generate the steam (state 10), from the water at ambient temperature (state 9). The exhaust gases (state 8), as an energy source, enters the generator where it separates the solution resulting in an H_2O-LiBr mixture with a low concentration (state 11) and the generation of refrigerant saturated steam (state 12). The mixture is modeled as a sub-cooled liquid type (state 19), which expands through a flow valve and arrives at the absorber as a low concentration H_2O-LiBr mixture (state 20). On the other hand, the heat is removed inside the condenser heat exchanger from the refrigerant to the environment (state 23), going from a gaseous to a liquid phase (state 13).

In the evaporator heat exchanger, the fluid takes heat from the refrigerated space or room, and induces a phase change in the refrigerant producing a pressure difference between the evaporator and the absorber, where the refrigerant exits as saturated steam (state 15) directly to the absorber, in which there is an energy change between the external water (state 27) and lithium bromide (state 20). As a result, a loop of the lithium bromide mixture is obtained to give a saturated liquid solution (state 16). The pressure of this solution is increased and entered into the heat exchanger by the flow energy supplied by the motor of the pump (state 17) through a counter-current configuration, which increases the temperature to improve efficiency. Finally, the fluid arrives at the generator to continue with the system cycle (state 18).

2.2. Thermodynamic Modeling

In the thermodynamic modeling of the trigeneration system [33], the components of the system are considered as open systems where a steady-state mass balance is applied according to Equation (1). For the case of constant flow systems, such as the generator and absorber, this balance results as shown in Equation (2).

$$\sum \dot{m}_{out} = \sum \dot{m}_{in} \tag{1}$$

$$\sum \dot{m}_{out} \cdot x_{out} = \sum \dot{m}_{in} \cdot x_{in} \tag{2}$$

where x is the concentration, $\dot{m}_{out}$ and $\dot{m}_{in}$ are the output and input mass flows to the system in kg/s.

Also, the energy balance applied to each component of the trigeneration system based on the first law of thermodynamics is expressed in Equation (3).

$$\sum \dot{Q} - \sum \dot{W} = \sum \dot{m}_{out} \cdot h_{out} - \sum \dot{m}_{in} \cdot h_{in} \tag{3}$$

where h is the specific enthalpy in kJ/kg, $\dot{Q}$ is heat flow rate in kW, and $\dot{W}$ is the power rate in kW.

The performance coefficient of the ARS (COP_{ARS}) is expressed by Equation (4), which is defined as the ratio of the heat transfer of the evaporator ($\dot{Q}_{Evaporator}$) in kW, and the amount of heat transfer in the generator ($\dot{Q}_{Gener}$) plus the energy rate of the pump ($\dot{W}_P$), both in kW.

$$COP_{ARS} = \frac{\dot{Q}_{Evaporator}}{\dot{Q}_{Gener} + \dot{W}_P} \tag{4}$$

Applying the energy balance to each of the components of the trigeneration system gives the equations shown in Table 1.

Table 1. Energy balance equations by components of the trigeneration system.

Component	Energy Balance
Compressor	$\dot{m}_1 \cdot h_1 + \dot{W}_{comp} - \dot{m}_2 \cdot h_2 = 0$
Combustion Chamber	$(\dot{m}_3 \cdot h_3 + \dot{m}_4 \cdot h_4) \cdot n_{cc} - \dot{m}_5 \cdot h_5 = 0$
Turbine	$\dot{m}_5 \cdot h_5 - \dot{W}_{turb} - \dot{m}_6 \cdot h_6 = 0$
Pre-heater	$\dot{m}_7 \cdot h_7 - \dot{m}_6 \cdot h_6 = \dot{m}_3 \cdot h_3 - \dot{m}_2 \cdot h_2$
HRSG	$\dot{m}_7 \cdot h_7 - \dot{Q}_{HRSG} + \dot{m}_8 \cdot h_8 = 0$
Generator	$\dot{m}_{18} \cdot h_{18} + \dot{Q}_{gener} - \dot{m}_{12} \cdot h_{12} - \dot{m}_{11} \cdot h_{11} = 0$
Condenser	$\dot{m}_{12} \cdot h_{12} - \dot{m}_{13} \cdot h_{13} + \dot{Q}_{cond} = 0$
Evaporator	$\dot{m}_{13} \cdot h_{13} + \dot{Q}_{evap} - \dot{m}_{15} \cdot h_{15} = 0$
Absorber	$\dot{m}_{15} \cdot h_{15} + \dot{m}_{20} \cdot h_{20} - \dot{m}_{16} \cdot h_{16} + \dot{Q}_{abs} = 0$
Heat exchanger	$\dot{m}_{17} \cdot h_{17} + \dot{m}_{11} \cdot h_{11} - \dot{m}_{18} \cdot h_{18} + \dot{m}_{20} \cdot h_{20} = 0$

To calculate the specific physical exergy ($\dot{E}^{Ph}$) was not considered the kinetic and potential energy, resulting in the Equation (5).

$$\dot{E}^{Ph} = (h - h_0) - T_0 \cdot (s - s_0) \tag{5}$$

where h is the specific enthalpy in kJ/kg, s is the specific entropy in kJ/kg · K of the working fluid flow, h_0 and s_0 are the state enthalpy and entropy at reference condition ($T_0 = 298.15$ K and $P_0 = 101.325$ kPa).

On the other hand, the chemical exergy for water ($\dot{E}_{water}^{Ch}$) was calculated using Equation (6), while for the microturbine exhaust gases (states 6, 7, 8, and 21) was used the Equation (6) since the change of chemical exergy for lithium bromide was not considered.

$$\dot{E}_{water}^{Ch} = \dot{m} \cdot \left(\frac{z_{water}}{M_{water}}\right) \cdot \dot{E}_{Ch,\,water}^{0} \tag{6}$$

$$\dot{E}^{ch} = \sum_{k=1}^{n} x_k \cdot \dot{E}^{ch_k} + R \cdot T_0 \sum_{k=1}^{n} x_k \cdot \ln x_k \tag{7}$$

where ($\dot{E}_{Q,\,water}^{0}$) is the standard chemical exergy of the water, x_k is the molar fraction, and ex^{ch_k} is the exergy per mol unit for the k gas.

The exergy balance was applied to each component of the trigeneration system according to Equation (8) [34].

$$\sum \dot{m}_{in} \cdot \dot{E}_{in} - \sum \dot{m}_{out} \cdot \dot{E}_{out} + \dot{Q} \cdot \left(1 - \frac{T_0}{T}\right) - \dot{W} - \dot{E}_D = 0 \tag{8}$$

where $\dot{m}_{in} \cdot \dot{E}_{in}$ is the inflow exergy, $\dot{m}_{out} \cdot \dot{E}_{out}$ is the outflow exergy, and $\dot{E}_D$ is the destroyed exergy.

The exergetic efficiency (η_{ex}) based on the second law of thermodynamics, is expressed by the Equation (9).

$$\eta_{ex} = \frac{\dot{E}_P}{\dot{E}_F} \tag{9}$$

where the amount of fuel exergy ($\dot{E}_F$) to the system, and the exergy produced ($\dot{E}_P$) per system are related to the destroyed exergy ($\dot{E}_D$), and the lost exergy ($\dot{E}_L$), as shown in Equation (10).

$$\dot{E}_F = \dot{E}_P + \dot{E}_D + \dot{E}_L \tag{10}$$

The Fuel and Product structure in each component of the trigeneration system was calculated, as shown in Table 2.

Table 2. Fuel and Product exergy equations.

Component	$\dot{E}_F$	$\dot{E}_P$	$\dot{E}_L$
Compressor	$\dot{W}_{comp}$	$\dot{E}_2 - \dot{E}_1$	-
Air pre-heater	$\dot{E}_6 - \dot{E}_7$	$\dot{E}_3 - \dot{E}_2$	-
Combustion Chamber	$\dot{E}_4$	$\dot{E}_5 - \dot{E}_3$	-
Turbine	$\dot{E}_5 - \dot{E}_6$	$\dot{W}_{turb}$	-
HRSG	$\dot{E}_7 - \dot{E}_8$	$\dot{E}_9 - \dot{E}_{10}$	-
Generator	$\dot{E}_8 - \dot{E}_{21}$	$\dot{E}_{12} + \dot{E}_{11} - \dot{E}_{18}$	$\dot{E}_{21}$
Condenser	-	-	$\dot{E}_{23}$
Evaporator	$\dot{E}_{14} - \dot{E}_{15}$	$\dot{E}_{24} - \dot{E}_{25}$	-
Absorber	$\dot{E}_{16} - \dot{E}_{15} - \dot{E}_{20}$	$\dot{E}_{27} - \dot{E}_{26}$	-
Heat exchanger	$\dot{E}_{11} - \dot{E}_{19}$	$\dot{E}_{18} - \dot{E}_{17}$	-

2.3. Thermo-Economic Analysis

To calculate the total production cost, it is considered the capital investment costs ($\dot{Z}_{CI}$), operation and maintenance ($\dot{Z}_{OM}$), as shown in Equation (11).

$$\dot{Z} = \dot{Z}_{CI} + \dot{Z}_{OM} \tag{11}$$

The equations used to calculate the Purchase Equipment Costs (PEC) for the components of the ARS were: heat exchangers (Equation (12)), pump (Equation (13)), motor (Equation (14)), where the sub-index "0" represents the reference of the studied component [35–37].

$$PEC_K = PEC_{0,K} \cdot \left(\frac{A_k}{A_0}\right)^{0.6} \quad (12)$$

where the reference area (A_0) is 100 m^2, the reference costs ($PEC_{0,K}$) considered are Evaporator (16,000 USD), Condenser (8000 USD), Absorber (16,500 USD), and Heat Exchanger (12,000 USD) [26]. Also, the PEC for the pump is calculated based on Equation (13).

$$PEC_{pump} = PEC_{0,pump} \cdot \left(\frac{\dot{W}_{pump}}{\dot{W}_{0,pump}}\right)^{m_B} \cdot \left(\frac{1-\eta_{pump}}{\eta_{pump}}\right)^{n_{pump}} \quad (13)$$

where the pump efficiency (η_{pump}) is 75%, the pump size power ratio (m_B) is 0.26, and the reference cost ($PEC_{0,pump}$) is 2100USD. In addition, the model used to estimate the PEC of the pump motor is presented in Equation (14).

$$PEC_{mot} = PEC_{0,mot} \cdot \left(\frac{\dot{W}_{mot}}{\dot{W}_{0,mot}}\right)^{m_{mot}} \cdot \left(\frac{1-\eta_{mot}}{\eta_{mot}}\right)^{n_{mot}} \quad (14)$$

where the motor size power ratio (m_{mot}) is 0.87, the motor reference power ($\dot{W}_{0,mot}$) is 10 kW, the motor efficiency (η_{mot}) is 90%, the efficiency ratio of motor size (n_{mot}) is 1, and the reference cost ($PEC_{0,mot}$) is 500 USD [26].

The components of the gas microturbine were used some well-known models [26]. For the PEC of the compressor was used the Equation (15), combustion chamber (Equation (16)), and turbine (Equation (17)).

$$PEC_{comp} = \left(\frac{C_{11} \cdot \dot{m}_{air}}{C_{12} - n_{comp}}\right) \cdot \left(\frac{P_{a2}}{P_{a1}}\right) \cdot \ln\left(\frac{P_{a2}}{P_{a1}}\right) \quad (15)$$

where the compressor coefficients C_{11} and C_{12} are 71.10 and 0.9 USD/(kg/s), respectively. In addition, the PEC of the combustion chamber was calculated according to Equation 16.

$$PEC_{cc} = \left(\frac{C_{21} \cdot \dot{m}_{CH4}}{C_{22} - \frac{P_{gc4}}{P_{a3}}}\right) \cdot \left[1 + e^{(C_{23} \cdot T_4 - C_{24})}\right] \quad (16)$$

where C_{21} is 46.08 USD/(kg/s), C_{22} is 0.995, C_{23} is 0.018 K^{-1} and C_{24} is 26.4 [26]. Also, for the case of the turbine, Equation (17) was used to calculate the PEC, which is a relevant cost of the microturbine equipment.

$$PEC_{turb} = \left(\frac{C_{31} \cdot \dot{m}_{comb}}{C_{32} - n_{turb}}\right) \cdot \ln\left(\frac{P_{gc4}}{P_{gc5}}\right)\left[1 + e^{(C_{33} \cdot T_4 - C_{34})}\right] \quad (17)$$

where the model turbine coefficients are, C_{31} in 479.34 USD/(kg/s), C_{32} is 0.92, C_{33} is 0.036 K^{-1} and C_{34} is 54.4 [26].

On the other hand, the values of the leveled costs (PEC_L) by components were calculated by the mean of the Equation (18), which consider the money transactions occur at the end of each year in the economic life of the trigeneration system.

$$PEC_L = CRF \cdot \sum_{j=1}^{n} \frac{PEC_J}{(1+i_{eff})^j} \quad (18)$$

where CRF is the Capital Return Factor, i_{eff} is the interest rate and PEC_j is the value of the Purchase Equipment Costs in the year j. Also, the CRF is calculated using the Equation (19).

$$CRF = \frac{i_{eff}(1+i_{eff})^n}{(1+i_{eff})^n - 1} \qquad (19)$$

where n is the lifetime of the equipment.

To obtain the value of the capital investment in term of unit cost per time ($\dot{Z}_k$), without having to calculate leveled costs, the Equation (20) can be used.

$$\dot{Z}_k = \frac{PEC_k \cdot CRF \cdot \varphi}{\tau} \qquad (20)$$

where τ is the total operation time in hours of the system at full load, and φ is the maintenance factor [38].

2.4. Exergy Cost Balance and Thermo-Economic Indicators

The exergetic cost balance, as shown in Equation (21), the inefficiencies presented in the equipment are evaluated, and the intermediate and final cost of the streams of the thermal process. This analysis allowed us to estimate all exergies of stream in the trigeneration cycle, considering the total costs (acquisition costs, operating costs, and maintenance costs) [39–41].

$$\sum_{i=1}^{n} \dot{C}_{out,i} + \dot{C}_{W,i} = \sum_{i=1}^{n} \dot{C}_{in,i} + \dot{C}_{Q,i} + \dot{Z}_k \qquad (21)$$

where the terms $\dot{C}_{in,i}$ and $\dot{C}_{out,i}$ are the exergy costs associated with the inlet and outlet flow, which are calculated using the Equation (22). The terms $\dot{C}_{W,i}$ and $\dot{C}_{Q,i}$ are the costs associated with the power and heat transfer exergy cost, which are calculated by the mean of the Equations (23) and (24), respectively.

$$\dot{C}_i = c_i \cdot \dot{E}_i \qquad (22)$$

$$\dot{C}_{W,i} = c_W \cdot \dot{W} \qquad (23)$$

$$\dot{C}_{Q,i} = c_Q \cdot \dot{E}_Q \qquad (24)$$

where c_i, c_W and c_Q are the specific costs per unit of exergy expressed in dollars per Gigajoules (USD/GJ).

The equations presented in Table 3 are obtained, applying the general cost balances to each component of the trigeneration system. The cost balance can be expressed as shown in Equation (25), as a function of the cost rates of the exergy lost ($\dot{C}_{L,i}$), product cost rate ($\dot{C}_{P,i}$) according to Equation (26), fuel cost rate ($\dot{C}_{F,i}$) attending to Equation (27), and the cost rate of exergy destruction ($\dot{C}_{D,i}$) by means of Equation (28).

$$\dot{C}_{P,i} = \dot{C}_{F,i} - \dot{C}_{L,i} + \dot{Z}_i \qquad (25)$$

$$c_{P,i} = \frac{\dot{C}_{P,i}}{\dot{E}_{P,i}} \qquad (26)$$

$$c_{F,i} = \frac{\dot{C}_{F,i}}{\dot{E}_{F,i}} \qquad (27)$$

$$\dot{C}_{D,i} = c_{F,i} \dot{E}_{D,i} \qquad (28)$$

Table 3. Cost balance equation by components.

Component	Cost Balance Equations	Auxiliary Equations
Compressor	$\dot{C}_1 + \dot{C}_{comp} + \dot{Z}_{comp} = \dot{C}_2$	$\dot{C}_1 = 0$
Air preheater	$\dot{C}_2 + \dot{C}_6 + \dot{Z}_{ph} = \dot{C}_3 + \dot{C}_7$	$\frac{\dot{C}_7}{\dot{E}_7} = \frac{\dot{C}_7}{\dot{E}_6}$
Combustion chamber	$\dot{C}_3 + \dot{C}_4 + \dot{Z}_{cc} = \dot{C}_5$	$\dot{C}_4 = 139.18$
Turbine	$\dot{C}_5 + \dot{Z}_{turb} = \dot{C}_6 + \dot{C}_{comp} + \dot{C}_{turb}$	$\frac{\dot{C}_6}{\dot{E}_6} = \frac{\dot{C}_5}{\dot{E}_5}, \frac{\dot{C}_{comp}}{\dot{W}_{comp}} = \frac{\dot{C}_{turb}}{\dot{W}_{turb}}$
HRSG	$\dot{C}_7 + \dot{C}_{10} + \dot{Z}_{HRSG} = \dot{C}_8 + \dot{C}_9$	$\frac{\dot{C}_8}{\dot{E}_8} = \frac{\dot{C}_7}{\dot{E}_7}$, $\dot{C}_{10} = 0$
Generator	$\dot{C}_{gen} + \dot{C}_{18} + \dot{Z}_{gen} = \dot{C}_{12} + \dot{C}_{11}$ $\frac{\dot{C}_{12}}{\dot{m}_{water}(ex_{12}-ex_{18})} - \frac{\dot{C}_{18}(ex_{11}-ex_{12})}{\dot{m}_t(ex_{12}-ex_{18})(ex_{11}-ex_{18})} - \frac{\dot{C}_{11}}{\dot{m}_{pump}(ex_{11}-ex_{18})} = 0$	-
Heat exchanger	$\dot{C}_{17} + \dot{C}_{11} + \dot{Z}_{he} = \dot{C}_{18} + \dot{C}_{19}$	$\frac{\dot{C}_{11}}{\dot{E}_{11}} = \frac{\dot{C}_{19}}{\dot{E}_{19}}$
Pump	$\dot{C}_{pump} + \dot{C}_{16} + \dot{Z}_{pump} = \dot{C}_{17}$	-
Condenser	$\dot{C}_{12} + \dot{Z}_{cond} = \dot{C}_{13} + \Delta \dot{C}_{cond}$	$\frac{\dot{C}_{12}}{\dot{E}_{12}} = \frac{\dot{C}_{13}}{\dot{E}_{13}}$
Absorber	$\dot{C}_{20} + \dot{C}_{15} + \dot{Z}_{abs} = \dot{C}_{16} + \Delta \dot{C}_{abs}$	$\frac{\dot{C}_{20}+\dot{C}_{15}}{\dot{E}_{20}+\dot{E}_{15}} = \frac{\dot{C}_{16}}{\dot{E}_{16}}$
Evaporator	-	$\frac{\dot{C}_{14}}{\dot{E}_{14}} = \frac{\dot{C}_{15}}{\dot{E}_{15}}$
Solution expansion valve	-	$\frac{\dot{C}_{19}}{\dot{E}_{19}} = \frac{\dot{C}_{20}}{\dot{E}_{20}}$
Coolant expansion valve	-	$\frac{\dot{C}_{13}}{\dot{E}_{13}} = \frac{\dot{C}_{14}}{\dot{E}_{14}}$

The relative cost difference (r) is calculated from the specific fuel and product cost, which shows the average relative cost increase per exergetic unit between the input power and the product, as shown in Equation (29).

$$r = \frac{c_{P,i} - c_{F,i}}{c_{F,i}} \quad (29)$$

Likewise, another thermo-economic indicator studied is the exergo-economic factor (f), which measures the relation between the capital cost investment compared to the loss costs rate and exergy destruction, and it is calculated using Equation (30).

$$f = \frac{\dot{Z}_i}{\dot{Z}_i + \dot{C}_{L,i} + \dot{C}_{D,i}} \quad (30)$$

A high value of this factor reflects a decrease in investment costs by improving energy efficiency, in contrast to low rates of this factor, which suggests saving costs throughout the system for improvement in efficiency.

3. Results and Discussion

This section presents the results of the energetic, exergetic, and thermo-economic analysis of the base case of the trigeneration system, and the parametric study result at different inlet air compressor temperatures. Table 4 presents the thermodynamic properties, physical and chemical exergy of the trigeneration system obtained from the energy balance.

Table 4. Thermodynamic properties, physical and chemical exergy of the trigeneration system.

State	$\dot{m}$ [kg/s]	T [K]	P [bar]	h [kJ/kg]	s [kJ/kg·K]	$\dot{E}^{Ph}$ [kJ/s]	$\dot{E}^{ch}$ [kJ/s]
1	0.30	298.15	1.013	298.60	5.69	0	0
2	0.30	452.10	3.64	454.31	6.16	46.71	0
3	0.30	673.20	3.61	684.83	6.16	73.83	0
4	0.50	298.15	3.70	−3.62	−0.68	99.72	2.5919
5	0.80	829.10	3.61	216.40	7.69	305.40	4.24
6	0.80	745.20	1.081	120.81	7.92	174.60	4.24
7	0.80	540	1.051	−108.10	7.57	75.11	4.24
8	0.80	346.50	1.021	−314.90	7.10	20.54	4.24
9	0.065	372.30	0.98	2651	7.36	29.85	617.50
10	0.065	298.15	1.013	104.81	0.36	0	0
11	0.57	346.50	0.059	167.91	0.44	23.60	0.62
12	0.019	346.50	0.059	2637	8.54	1.84	0.049
13	0.019	309.20	0.059	150.81	0.52	0.014	0.049
14	0.019	278.20	0.0087	150.80	0.54	−0.13	0.049
15	0.019	278.20	0.0087	2510	9.024	−3.49	0.049
16	0.59	307.20	0.0087	81.38	0.20	14.19	0.66
17	0.59	307.20	0.059	81.38	0.20	14.19	0.66
18	0.59	331.50	0.059	131.20	0.36	16.081	0.66
19	0.57	322.90	0.059	120.20	0.29	20.63	0.62
20	0.57	310.90	0.0087	120.20	0.22	33.51	0.62

These values allow conducting the exergy balance to calculate the exergy destruction in each component of the system, obtaining the values shown in Figure 2, which presents the percentage of the exergy destruction fraction by equipment and/or components of the system studied.

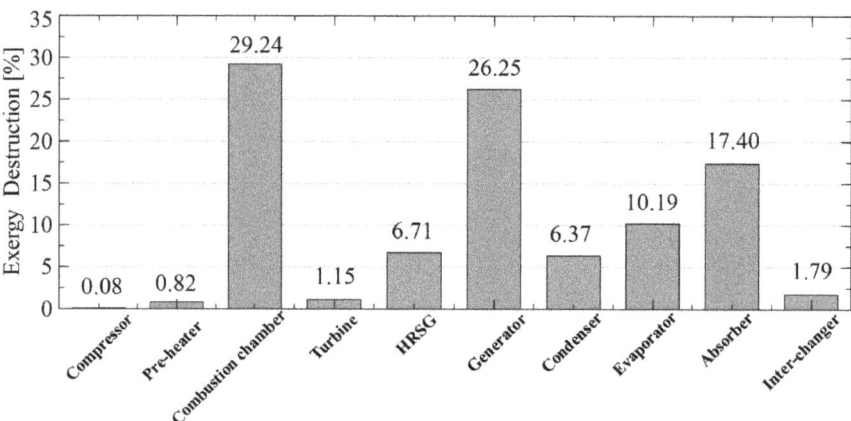

Figure 2. Percentage Exergy Destruction for each component of the trigeneration system.

The exergy destruction represents a loss of useful work that can be taken advantage of by the components of the trigeneration system for the improvement of the operating and thermo-economic conditions. This translates into a great inefficiency and a considerable quantity of energy that must be minimized just when it is needed to maximize the overall thermal efficiency of the process. However, it represents an opportunity to optimize and develop innovate techniques based on new proposals and alternatives of operation and manufacture to design and to rate the devices, all with the purpose of reducing the investment cost associated with the components and, therefore, to the general system.

The results show that the combustion chamber of the gas microturbine is the component with the greatest exergy destruction (29.24%), followed by the generator of the ARS (26.25%). The compressor has a contribution of 0.08% due to the low heat transfer irreversibility presented in this device as a result of the high operating temperature. The greatest amount of exergy destroyed (72%) is located in the combustion chamber, generator and absorber, which suggests a greater technological and operational effort focused on the design of these heat exchange, allowing us to reduce the temperature difference between the fluids. Likewise, from the component cost balances, the exergetic costs of the stream in the system were calculated, as shown in Table 5.

Table 5. Exergetic cost rates ($\dot{C}$) in USD/s, and the costs per unit of exergy of the stream (c) in USD/GJ.

State	$\dot{C}$ [10^{-3} USD/s]	c [USD/GJ]	State	$\dot{C}$ [10^{-3} USD/s]	c [USD/GJ]
1	0	0	11	3.18	25.73
2	89.53	532.4	12	0.42	24.27
3	198.50	746.9	13	0.014	24.27
4	139.21	1.48	14	0.017	24.27
5	338.70	303.88	15	0.75	24.27
6	195.61	303.87	16	1.39	26.14
7	86.81	303.89	17	1.48	27.80
8	27.12	303.88	18	1.86	30.89
9	60.67	303.90	19	2.85	25.73
10	0	0	20	3.61	25.73

These costs allow us to determine the cost of raw materials, products, and destruction, besides the thermo-economic indicators shown in Table 6. The highest costs are associated with the pre-heater assembly (532.40 USD/GJ), which exceeds the cost of the products obtained by the turbine.

Table 6. Average costs per unit of fuel ($\dot{C}_{F,i}$) and product ($\dot{C}_{P,i}$), destruction cost ($\dot{C}_{D,i}$), relative cost difference (r), and exergo-economic factor (f) for each component of the trigeneration system.

Equipment	$\dot{C}_{F,i}$ [USD/s]	$\dot{C}_{P,i}$ [USD/s]	$\dot{C}_{D,i}$ [USD/s]	r	f
Compressor	24.80	24.87	0	0.0031	1
Pre-heater	30.22	30.27	79.13	2.67	0.0024
Combustion chamber	38.66	38.95	137.91	110.20	0.0072
Turbine	39.75	15.80	110.50	0.75	0.027
HRSG	16.58	16.85	648.51	0.91	0.0015
Generator	12.94	22.031	0.43	0.70	0.41
Assembly evaporator	26.35	132.70	1.11	4.30	0.28
Heat exchanger	27.15	65.77	0.15	1.42	0.41

3.1. Energy and Exergy Analysis

This section presents the parametric study results applied in the trigeneration system, through the variation in compressor inlet temperature (T_1) from 293.15 K to 328.15 K. The energy performance was evaluated for the microturbine subsystem, HRSG, and the evaporator of the ARS, which will be analyzed and discussed in detail below.

Figure 3a shows that the increase in temperature causes a decrease in the net power supplied by the turbine with respect to the heat absorbed by the evaporator at different air flow ratio, because of the enhancement in the heat removed in the evaporator as a consequence of the increase in the air flow temperature in the inlet compressor. For a given inlet compressor temperature of 313.15 K, the Trigeneration System delivers almost 236.2% power per unit of heat in the evaporator with an air-fuel ratio of 0.7 with respect to 0.5. This means that an increase in the air-fuel ratio causes the microturbine to deliver more power, given the higher airflow. However, the decreasing trend among the studied energy coefficient is preserved, so the inlet compressor temperature is a basic parameter that significantly affects the energy performance of the trigeneration system. On the other hand, the steam energy supplied by the HRGS (Figure 3b) increases with the increase of the air flow inlet compressor temperature, which is due to the improvement of the combustion chamber thermal efficiency, having a combustion with an oxidant both at higher temperature, and more air mass flow by increasing the air-fuel ratio, which increases the thermal capacitance in the HRSG.

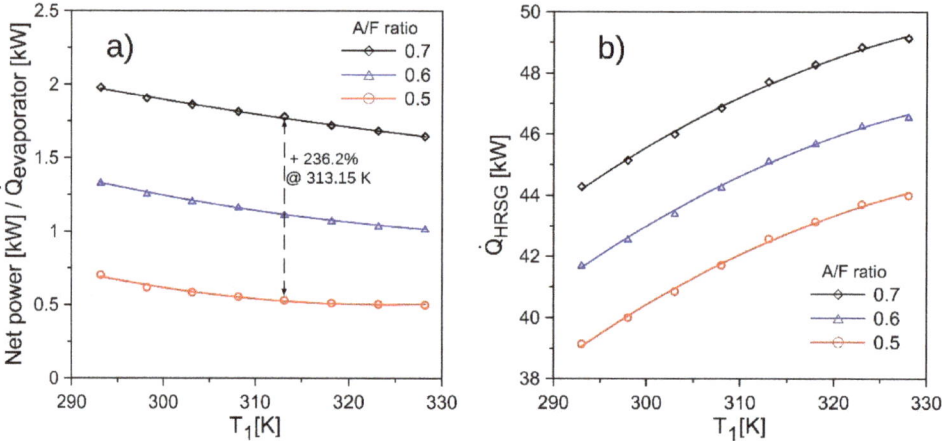

Figure 3. Energy performance of the trigeneration system at different compressor inlet air temperatures, (**a**) Net Power/Evaporator heat, and (**b**) Heat in the Heat Recovery Steam Generator (HRSG).

3.2. Exergy Destruction

The exergy destruction analysis by component represents an opportunity for the improvement of the operative and thermo-economic of the thermal system. With this parameter, some alternatives can be developed to reduce the effects of both economic and operational losses. It is important to know the behavior of the minimum and maximum values that can be presented in a device with the purpose of obtaining the possible exergy losses in the trigeneration system and the effect of the increase of the inlet air compressor temperature in the devices of the system that is studied here. The exergy destroyed in the trigeneration system at different compressor inlet air temperatures is shown in Figure 4.

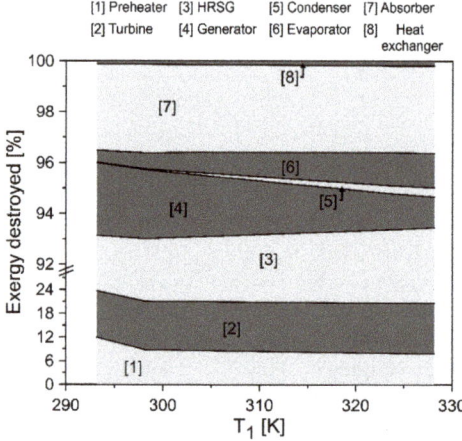

Figure 4. Exergy destroyed in the trigeneration system at different compressor inlet air temperatures.

For the exergy destroyed fraction in terms of components, a tendency was observed. We used an accumulation of the percentage losses to obtain a graphical model that describes the behavior of the exergy losses of the general system. With the exclusion of the combustion chamber in the study, we observed a 12% reduction in the exergy destroyed in the turbine until 300 K. Then, it remained at its lowest values while the inlet air compressor temperature increased. Another relevant result is the behavior of the exergy destruction in the generator and condenser, which both present the same range of variation of exergy destroyed (4% to 5%). In this case, below the temperature of 300 K, the exergy destroyed in the generator is greater than the values presented. After this, an increase in temperature caused a significant enhancement of the exergy efficiency of this device at higher compressor inlet air temperature. Therefore, the highest values of exergy destroyed in the devices of the trigeneration system happen when the air temperature is lower than 300 K. However, there are exceptions such as the HRSG and the evaporator, because a higher air temperature limits their functionality in the system.

3.3. Cost Rate of Exergy Destruction

The costs related to the exergy destruction represent the economic losses in dollars with respect to the operating time of the system and, therefore, of each one of the components. A comparative analysis of the costs per component of the microturbine is proposed for the variation of the inlet air compressor temperature, while simultaneously analyzing the behavior of the costs related to the exergy destruction in the assembly and different components both in the ARS and microturbine. Figure 5 presents the trends of the cost rate of exergy destruction of the trigeneration system at different compressor inlet air temperatures.

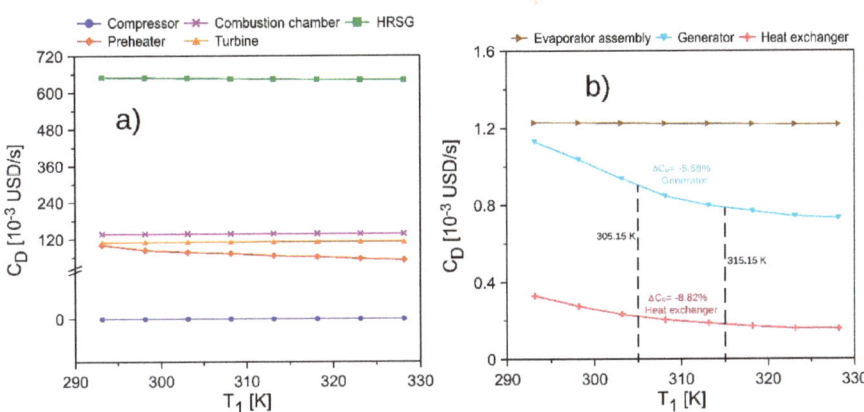

Figure 5. Cost rate of exergy destruction of the trigeneration system at different compressor inlet air temperatures, (**a**) microturbine components, and (**b**) absorption chiller components.

The results show that for the ARS (Figure 5b), both the generator and the heat exchanger decrease their costs as the temperature increases, while in the evaporator assembly there is an unusual behavior which suggests that the variation in the compressor air temperature does not infer exergy destruction or, therefore, related costs. In addition, it is evident that both the evaporator assembly and the heat exchanger have lower costs in the microturbine than in the ARS, which can be explained by the operating conditions of the refrigerant fluids, which directly influence the efficiency of the general system.

3.4. Relative Cost Difference and Exergo-Economic Factor

The exergo-economic factor results evidenced that, in the systems, there are values of 100% for the compressor and this remains constant for changes of temperature as shown in Figure 6 because the flow energy input in microturbine is free and the destroyed exergy cost rate of the compressor is equal to zero.

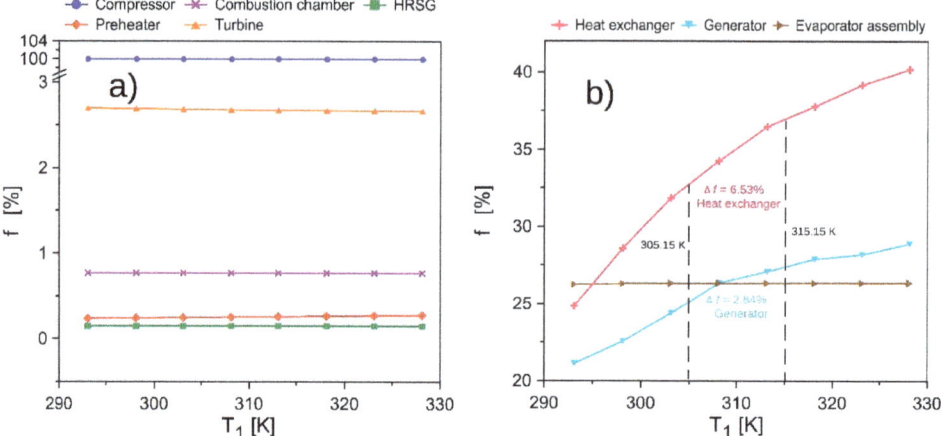

Figure 6. Exergoeconomic factor of the trigeneration system at different compressor inlet air temperatures, (**a**) microturbine components, and (**b**) absorption chiller components.

Therefore, in the microturbine, the compressor is the equipment in the microturbine with the lowest exergo-economic values, which implies that destruction costs and maintenance costs are relevant in the system. On the other hand, in the ARS system, both the generator and the heat exchanger have high values compared to the microturbine system, which is the same in the relative cost behavior of the evaporator assembly with a very high exergo-economic factor of around 26%. Thus, the ARS has a higher exergo-economic factor than the microturbine system, which is a consequence of high acquisition costs, which are more relevant than the costs for exergy destruction and maintenance, which suggests a reduction of non-fixed costs that can be modified.

In this case, the relative costs increase as the compressor air temperature increases, which indicates a thermodynamic limitation in the system due to the temperature limits allowed. However, a temperature increase of 10 K provides improvements in the overall performance of the microturbine components, while the other subsystem suffers increases of 6.53% and 2.84% of the exergo-economic factor in the heat exchanger and generator, respectively, as shown in Figure 6b.

Figure 7 enables the analysis of the relative cost difference of the main components of the trigeneration systems in this study. Thus, we obtained the tendency of this thermo-economic indicator in a wide range of operations, which allows us to determine the critical equipment that represents the main construction cost of the system and, subsequently, to reduce these costs through design strategies or operational changes of the general thermal system.

In the microturbine subsystem (Figure 7a), the results show a very little effect of the inlet air compressor temperature on the relative costs associated with the different components, which is a consequence of the law variation presented in the exergy destroyed by component and similar values of product and fuel cost in the range of the evaluated temperature.

The results show that, in the case of the ARS, the variation of the compressor inlet temperature causes a decrease in the relative cost of the evaporator assembly (21.63%) with increasing temperature from 305.15 K to 315.15 K, as shown in Figure 7b. However, this behavior does not occur in the entire range studied, which allows us to predict the thermo-economic indicator of the trigeneration system under different ambient temperatures.

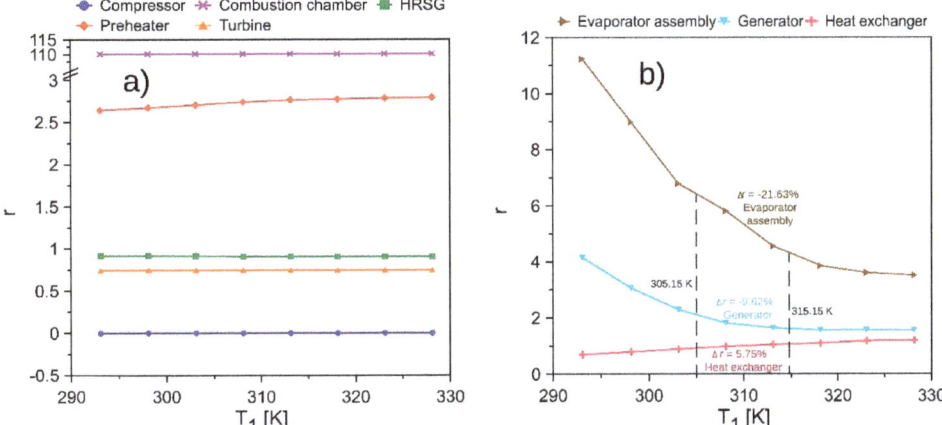

Figure 7. Relative cost difference of the trigeneration system at different compressor inlet air temperatures, (**a**) Microturbine components, and (**b**) Absorption chiller components.

4. Conclusions

This study has been carried out considering a trigeneration system with defined limits and considerations. The modification of these considerations and extension of the limits of this process will require an evaluation of some exergetic costs in the exergo-economic model that were not considered, in addition to the exergy required in the condenser, evaporator and absorber currents. The exergetic costs not considered, if evaluated, could compromise the thermo-economic viability of the system, which must be evaluated in detail by means of thermo-economic indicators such as the recovery period of the investment, the specific investment cost and the leveled cost of the energy since this was not defined in the scope of the present study.

The study allowed us to undertake a thermo-economics approach to evaluate energy conversion systems from the energy perspective, and in a broad way, in complement with economic considerations. It also enabled us to verify the real viability of the trigeneration system when it operates with different air temperatures at the inlet of the compressor.

The thermo-economic parametric was developed based on a gas microturbine-ARS–HRSG trigeneration system model, evaluating the different exergy destroyed cost, exergo-economic factor, and relative cost according to each component. The result also showed the opportunities for improvement in components, and the amount of useful energy available that can be recovered from the exhaust gases of the gas microturbine. However, there is an operation condition where the exergy is highly destroyed due to the exergy inefficiencies of the equipment, and a greater purchase equipment cost is presented with a high exergo-economic factor.

In addition, it is concluded that the system is technically and economically viable, which represents a considerable alternative for the implementation in the secondary energy sector, where the cooling and power generation is required, such as the operation of shopping malls, supermarkets, and hotels.

Author Contributions: Conceptualization: G.V.O.; Methodology: G.V.O. and C.A.P.; Software: G.V.O., J.D.F. and C.A.P.; Validation: G.V.O., J.D.F. and C.A.P.; Formal Analysis: G.V.O., J.D.F. and C.A.P.; Investigation: G.V.O.; Resources: G.V.O. and C.A.P.; Writing-Original Draft Preparation: G.V.O.; Writing-Review & Editing: J.D.F. and C.A.P.; Funding Acquisition: G.V.O.

Funding: This work was supported by the Universidad del Atlántico and Universidad Francisco de Paula Santander.

Acknowledgments: The authors are grateful to Universidad del Atlántico, Universidad Francisco de Paula Santander, and Jhonatan De la Cruz for his collaboration received in the case study simulation.

Conflicts of Interest: The authors declare no conflict of interest.

Abbreviations

The following abbreviations are used in this manuscript:

HRSG	Heat Recovery Steam Generator
ARS	Absorption Refrigeration System
CRF	Capital Recovery Factor
DC	Direct cost
CI	Capital Investment
PEC	Purchase Equipment Cost
CC	Combustion Chamber
TCI	Total Capital Investment
OM	Operation and Maintenance

Nomenclature

c	Specific cost (USD/GJ)
s	Entropy (kJ/kgK)
$\dot{C}$	Associated cost (USD/s)
$\dot{E}$	Exergy (kJ/s)
h	Specific enthalpy (kJ/kg)
$\dot{m}$	Mass flow (kg/s)
$\dot{Q}$	Heat transfer (kJ/s)
R	Universal gas constant (atm L/mol K)
P	Pressure (bar)
rpm	Rotational engine speed (rpm)
T	Temperature (K)
t	Time (s)
$\dot{W}$	Power (kW)
X	fraction
A	Area (m^2)
COP	Performance coefficient
i_{eff}	Annual interest rate
n	Equipment's lifetime (year)
$\dot{Z}$	capital investment cost (USD/s)
r	Relative cost
f	The exergoeconomic factor

Greek Letters

φ	Maintenance factor
η	Heat recovery efficiency
τ	Total operation time (Hr)

Subscripts

ex	Exergetic
mot	Motor
k	Molar
$cond$	Condenser
abs	Absorber
$evap$	Evaporator
ch	Chemical
ph	Physical
o	Standard or reference
D	Destruction
L	Lost
P	Produced
k	Molar
F	Supplied
B	Pump size

References

1. Urbanucci, L.; Testi, D.; Bruno, J.C. Integration of reversible heat pumps in trigeneration systems for low-temperature renewable district heating and cooling microgrids. *Appl. Sci.* **2019**, *9*, 3194. [CrossRef]
2. Ramírez, R.; Gutiérrez, A.S.; Eras, J.J.C.; Valencia, K.; Hernández, B.; Forero, J.D. Evaluation of the energy recovery potential of thermoelectric generators in diesel engines. *J. Clean. Prod.* **2019**, *241*, 118412. [CrossRef]
3. Valencia, G.; Duarte, J.; Isaza-Roldan, C. Thermoeconomic analysis of different exhaust waste-heat recovery systems for natural gas engine based on ORC. *Appl. Sci.* **2019**, *9*, 4017. [CrossRef]
4. Al Moussawi, H.; Fardoun, F.; Louahlia-Gualous, H. Review of tri-generation technologies: Design evaluation, optimization, decision-making, and selection approach. *Energy Convers. Manag.* **2016**, *120*, 157–196. [CrossRef]
5. Lecompte, S.; Lemmens, S.; Huisseune, H.; Van den Broek, M.; De Paepe, M. Multi-objective thermo-economic optimization strategy for ORCs applied to subcritical and transcritical cycles for waste heat recovery. *Energies* **2015**, *8*, 2714–2741. [CrossRef]
6. Aprhornratana, S.; Eames, I. Thermodynamic analysis of absorption refrigeration cycles using the second law of thermodynamics method. *Int. J. Refrig.* **1995**, *18*, 244–252. [CrossRef]
7. Florides, G.A.; Kalogirou, S.A.; Tassou, S.A.; Wrobel, L.C. Modelling, simulation and warming impact assessment of a domestic-size absorption solar cooling system. *Appl. Therm. Eng.* **2002**, *22*, 1313–1325. [CrossRef]
8. Shirmohammadi, R.; Soltanieh, M.; Romeo, L. Thermoeconomic analysis and optimization of post-combustion CO_2 recovery unit utilizing absorption refrigeration system for a natural-gas-fired power plant. *Environ. Prog. Sustain. Energy* **2018**, *37*, 1075–1084. [CrossRef]
9. Valencia, G.; Núñez, J.; Duarte, J. Multiobjective optimization of a plate heat exchanger in a waste heat recovery organic rankine cycle system for natural gas engines. *Entropy* **2019**, *21*, 655. [CrossRef]
10. Szargut, J.; Morris, D.; Stewar, F. *Exergy Analysis of Thermal, Chemical and Metallurgical Processes*; Hemisphere Publishing: New York, NY, USA, 1988.
11. Kotas, T. *The Exergy Method of Thermal Plant Analysis*; Anchor Brendon Ltd.: London, UK, 1985.
12. Kaynakli, O.; Kilic, M. Theoretical study on the effect of operating conditions on performance of absorption refrigeration system. *Energy Convers. Manag.* **2007**, *48*, 599–607. [CrossRef]
13. Martínez, H.; Rivera, W. Energy and exergy analysis of a double absorption heat transformer operating with water/lithium bromide. *Int. J. Energy Res.* **2009**, *33*, 662–674. [CrossRef]
14. Kaushik, S.C.; Arora, A. Energy and exergy analysis of single effect and series flow double effect water–lithium bromide absorption refrigeration systems. *Int. J. Refrig.* **2009**, *32*, 1247–1258. [CrossRef]
15. Gomri, R.; Hakimi, R. Second law analysis of double effect vapour absorption cooler system. *Energy Convers. Manag.* **2008**, *49*, 3343–3348. [CrossRef]
16. Rosiek, S. Exergy analysis of a solar-assisted air-conditioning system: Case study in southern Spain. *Appl. Therm. Eng.* **2019**, *148*, 806–816. [CrossRef]
17. Pourfayaz, F.; Imani, M.; Mehrpooya, M.; Shirmohammadi, R. Process development and exergy analysis of a novel hybrid fuel cell-absorption refrigeration system utilizing nanofluid as the absorbent liquid. *Int. J. Refrig.* **2019**, *97*, 31–41. [CrossRef]
18. Alanne, K.; Saari, A. Sustainable small-scale CHP technologies for buildings: The basis for multi-perspective decision-making. *Renew. Sustain. Energy Rev.* **2004**, *8*, 401–431. [CrossRef]
19. Al-Sulaiman, F.A.; Hamdullahpur, F.; Dincer, I. Trigeneration: A comprehensive review based on prime movers. *Int. J. Energy Res.* **2011**, *35*, 233–258. [CrossRef]
20. Mohammadi, S.M.H.; Ameri, M. Energy and exergy analysis of a tri-generation water-cooled air conditioning system. *Energy Build.* **2013**, *67*, 453–462. [CrossRef]
21. Fontalvo, A.; Pinzon, H.; Duarte, J.; Bula, A.; Quiroga, A.G.; Padilla, R.V. Exergy analysis of a combined power and cooling cycle. *Appl. Therm. Eng.* **2013**, *60*, 164–171. [CrossRef]
22. Diaz, G.A.; Forero, J.D.; Garcia, J.; Rincon, A.; Fontalvo, A.; Bula, A.; Padilla, R.V. Maximum power from fluid flow by applying the first and second laws of thermodynamics. *J. Energy Resour. Technol.* **2017**, *139*, 032903. [CrossRef]
23. Marimón, M.A.; Arias, J.; Lundqvist, P.; Bruno, J.C.; Coronas, A. Integration of trigeneration in an indirect cascade refrigeration system in supermarkets. *Energy Build.* **2011**, *43*, 1427–1434. [CrossRef]

24. Hawkes, A.D.; Leach, M.A. Cost-effective operating strategy for residential micro-combined heat and power. *Energy* **2007**, *32*, 711–723. [CrossRef]
25. Valencia Ochoa, G.; Acevedo Peñaloza, C.; Duarte Forero, J. Thermoeconomic Optimization with PSO Algorithm of Waste Heat Recovery Systems Based on Organic Rankine Cycle System for a Natural Gas Engine. *Energies* **2019**, *12*, 4165. [CrossRef]
26. Bejan, A.; Tsatsaronis, G.; Moran, M. *Thermal Design and Optimization*; John Wiley & Sons: Hoboken, NJ, USA, 1995; ISBN 0471584673.
27. Kizilkan, Ö.; Sencan, A.; Kalogirou, S. Thermoeconomic optimization of a LiBr absorption refrigeration system. *Chem. Eng. Process.* **2007**, *46*, 1376–1384. [CrossRef]
28. Lian, Z.T.; Chua, K.J.; Chou, S.K. A thermoeconomic analysis of biomass energy for trigeneration. *Appl. Energy* **2010**, *87*, 84–95. [CrossRef]
29. Mago, P.J.; Hueffed, A.K. Evaluation of a turbine driven CCHP system for large office buildings under different operating strategies. *Energy Build.* **2010**, *42*, 1628–1636. [CrossRef]
30. Ghaebi, H.; Karimkashi, S.; Saidi, M.H. Integration of an absorption chiller in a total CHP site for utilizing its cooling production potential based on R-curve concept. *Int. J. Refrig.* **2012**, *35*, 1384–1392. [CrossRef]
31. Ahmadi, P.; Rosen, M.A.; Dincer, I. Greenhouse gas emission and exergo-environmental analyses of a trigeneration energy system. *Int. J. Greenh. Gas Control* **2011**, *5*, 1540–1549. [CrossRef]
32. Ghaebi, H.; Saidi, M.H.; Ahmadi, P. Exergoeconomic optimization of a trigeneration system for heating, cooling and power production purpose based on TRR method and using evolutionary algorithm. *Appl. Therm. Eng.* **2012**, *36*, 113–125. [CrossRef]
33. Arregoces, A.J.; De la Cruz, J.R.; Valencia, G.E. Estudio comparativo del desempeño energético, económico y ambiental de un sistema de trigeneración constituido por una microturbina de 30kW acoplada a una tecnología de calefacción y diferentes tecnologías de refrigeración. *Univ. Atl.* **2019**, *1*, 123.
34. Moran, M.J.; Saphiro, H.N.; Boettner, D.D.; Bailey, M.B. *Fundamentals of Engineering Thermodynamics*; Wiley: Hoboken, NJ, USA, 2011; ISBN 978-1-118-10801-7.
35. Garousi Farshi, L.; Mahmoudi, S.M.S.; Rosen, M.A. Exergoeconomic comparison of double effect and combined ejector-double effect absorption refrigeration systems. *Appl. Energy* **2013**, *103*, 700–711. [CrossRef]
36. Misra, R.D.; Sahoo, P.K.; Gupta, A. Thermoeconomic optimization of a LiBr/H$_2$O absorption chiller using structural method. *J. Energy Resour. Technol.* **2005**, *127*, 119–124. [CrossRef]
37. Boyaghchi, F.A.; Mahmoodnezhad, M.; Sabeti, V. Exergoeconomic analysis and optimization of a solar driven dual-evaporator vapor compression-absorption cascade refrigeration system using water/CuO nanofluid. *J. Clean. Prod.* **2016**, *139*, 970–985. [CrossRef]
38. Bilgen, S.; Kaygusuz, K. Second law (exergy) analysis of cogeneration system. *Energy Sources Part A* **2008**, *30*, 1267–1280. [CrossRef]
39. Valencia, G.; Fontalvo, A.; Cárdenas, Y.; Duarte, J.; Isaza, C. Energy and exergy analysis of different exhaust waste heat recovery systems for natural gas engine based on ORC. *Energies* **2019**, *12*, 2378. [CrossRef]
40. Ochoa, G.V.; Peñaloza, C.A.; Rojas, J.P. Thermoeconomic modelling and parametric study of a simple ORC for the recovery of waste heat in a 2 MW gas engine under different working fluids. *Appl. Sci.* **2019**, *9*, 4526. [CrossRef]
41. Valencia, G.; Benavides, A.; Cárdenas, Y. Economic and Environmental Multiobjective Optimization of a Wind–Solar–Fuel Cell Hybrid Energy System in the Colombian Caribbean Region. *Energies* **2019**, *12*, 2119. [CrossRef]

© 2019 by the authors. Licensee MDPI, Basel, Switzerland. This article is an open access article distributed under the terms and conditions of the Creative Commons Attribution (CC BY) license (http://creativecommons.org/licenses/by/4.0/).

Article

Advance Exergo-Economic Analysis of a Waste Heat Recovery System Using ORC for a Bottoming Natural Gas Engine

Guillermo Valencia Ochoa [1,*], Jhan Piero Rojas [2] and Jorge Duarte Forero [1]

[1] Programa de Ingeniería Mecánica, Universidad del Atlántico, Carrera 30 Número 8-49, Puerto Colombia, Barranquilla 080007, Colombia; jorgeduarte@mail.uniatlantico.edu.co
[2] Facultad de Ingeniería, Universidad Francisco de Paula Santander, Avenida Gran Colombia No. 12E-96, Cúcuta 540003, Colombia; jhanpierorojas@ufps.edu.co
* Correspondence: guillermoevalencia@mail.uniatlantico.edu.co; Tel.: +57-5-324-94-31

Received: 19 November 2019; Accepted: 2 January 2020; Published: 5 January 2020

Abstract: This manuscript presents an advanced exergo-economic analysis of a waste heat recovery system based on the organic Rankine cycle from the exhaust gases of an internal combustion engine. Different operating conditions were established in order to find the exergy destroyed values in the components and the desegregation of them, as well as the rate of fuel exergy, product exergy, and loss exergy. The component with the highest exergy destroyed values was heat exchanger 1, which is a shell and tube equipment with the highest mean temperature difference in the thermal cycle. However, the values of the fuel cost rate (47.85 USD/GJ) and the product cost rate (197.65 USD/GJ) revealed the organic fluid pump (pump 2) as the device with the main thermo-economic opportunity of improvement, with an exergo-economic factor greater than 91%. In addition, the component with the highest investment costs was the heat exchanger 1 with a value of 2.769 USD/h, which means advanced exergo-economic analysis is a powerful method to identify the correct allocation of the irreversibility and highest cost, and the real potential for improvement is not linked to the interaction between components but to the same component being studied.

Keywords: advanced exergo-economic analysis; waste heat recovery system; ORC; endogenous exergy; exogenous exergy

1. Introduction

In thermodynamics systems, the irreversibility in the components of the system produce exergy destruction, and the continuous improvement of the performance of energy conversion systems based on exergy analysis has been a priority among researchers in this field of study [1], from energy analysis of the thermal systems [2] and trigeneration systems [3]. However, from all the information available with these traditional analyses, there is no relevant development on the analysis carried out with this methods to make further improvements to the components of a system [4], so advanced analyses and economic analysis are proposed to facilitate the thermal and economic improvement of the system.

In recent years, new concepts of exergy, such as endogenous/exogenous and avoidable/unavoidable exergy destruction, have been employed to obtain relevant information for the identification of irreversibilities and thermodynamic inefficiencies in the systems [5]. In any thermodynamic system, the exergy destruction in the components can be generated in two ways. The first is due to the irreversibilities of the component under study, which is called endogenous exergy destruction, while the second is due to the irreversibilities of the other components that affect the component under study; this is called exogenous exergy destruction [6]. Thus, its optimization process will depend on the technical and economic limitations of the system, so there will only be a part of the exergy destruction and avoidable/unavoidable investment costs for each component. By uniting these two concepts, it is possible to identify improvements to the system [7].

Long et al. [8] evaluated the importance of the working fluid in the thermal performance of an organic Rankine cycle (ORC) by means of an external and internal exergetic analysis, and an optimization analysis based on a genetic algorithm with exergetic efficiency as an objective function. The results of the exergy analysis showed that the organic working fluid affects the exergetic efficiency of the cycle, with the opposite case in the internal part, where the efficiency did not present changes. The results of the optimization showed that the selection of the working fluid depends, to a greater degree, on the optimal evaporation temperature, which increases the exergetic efficiency of the cycle. Long et al. [9] performed an exergy analysis to evaluate the impacts of the evaporation pressure and ammonia fraction on the ammonia–water mixture of the system performance Kalina, obtaining that the evaporation pressure plays an important value in the internal and external exergetic efficiency. Additionally, optimal values are obtained from these in their ideal operation, as well as the ammonia fractions increasing the exergetic efficiency depending on the evaporation pressure. However, the exergetic efficiency of the cycle depends on the input temperature of the heat source, evaluating the impact of this parameter on internal and external exergetic efficiency.

Tian et al. [10] developed a techno-economic analysis of a system consisting of an ORC and an internal combustion engine operating with 235 kW diesel, in order to study the performance of 20 organic fluids. The results showed that the highest energy generated per unit of mass flow and the highest energy efficiency are obtained for refrigerant R-141b and refrigerant R-123, respectively. The study is limited to a single engine operating condition, and a traditional exergetic analysis where the real opportunities for both endogenous and exogenous component savings are not shown.

On the other hand, Zare V. [11], in order to find savings opportunities, added economic criteria to the thermal performance studies, applied to three configurations of an ORC. However, this application was limited to binary geothermal power plants, where the RORC presented better energy results, while from the economic point of view, the simple ORC was the best option because it is integrated by a smaller amount of equipment, which implies a lower acquisition cost. The results do not consider the evaluation of costs by components but of a global system. In addition, studies from the exergetic point of view have been developed in a traditional way, and thermal-economical studies for waste heat recovery systems of gas generation engines through ORC have not been widely integrated. Thus, the literature reports the results of the modeling developed by Kerme and Orfi [12], who studied the effect of the temperature of the organic fluid at the entrance of an ORC turbine on the energy and exergy performance, obtaining that the increase of the temperature increases the efficiency while total exergy destruction decreases it.

The combination of traditional and advanced exergetic analysis can provide significant thermodynamic information, such as the source and the amount of exergy destroyed by each component [13], and how much this destruction can be avoided [14], as in the case of solar energy collectors with a flat plate and a flat plate with a thin plate, resulting in the exergy destruction in the absorbent plate being greater than the rest of components, but according to the advanced exergetic analysis performed, this exergy destruction is endogenous and unavoidable, which means that the irreversibilities of this component are inherent in its operation mode [15].

Mohammadi et al. [16] studied a combination of conventional and advanced exergetic analyses in a supercritical CO_2 recompression cycle to determine the potential for improving the thermal cycle performance, where the overall exergetic efficiency reached 17.13%, the system's maximum best potential was 106.85 MW, and approximately 35% of exergy destruction could be avoided by focusing on components, such as the heat exchanger, turbine, and main compressor. These investigations can be complemented with the help of the combination of exergetic analysis [17] and economic analysis to obtain thermo-economic costs based on the irreversibilities of the components [18].

In addition, comparative studies have been carried out on different configurations of waste heat recovery cycles integrated to gas engines [18] and applications, such as Petrakopoulou et al. [19], where the first application of an exergo-economic analysis in a CO_2 capture power plant was evaluated, revealing that the costs associated with exergy and investment analyses are endogenous for most components, where it proposed a suggestion for improving some components, such as the reactor, expander, and compressor. The literature review shows the case of a polygeneration plant operating in a geothermal cascade system coupled to an organic Rankine cycle that produces 40 kWe, where improvement potentials were found in the ORC cycle (10.61 kW) and heat exchanger (2.28 kW), while the exergo-economic analysis revealed an electricity production cost of 7.78 $/h and the advanced exergo-economic analysis suggests that the plant heat exchanger is the component with the greatest opportunity to reduce the exergy destruction of the heat exchanger equipment [20].

Another application was in a combined steam-organic Rankine cycle to recover waste heat from a gas turbine, where an exergo-economic analysis was performed using three different organic fluids (R124, R152a, and R34a), obtaining that the maximum exergy efficiency and the minimum rate of product costs are 57.62% and 396 $/h, respectively. In addition, the parametric study was complemented with genetic algorithm optimization, where it was obtained that the combined cycle with R152a has the best performance from the thermodynamic and exergo-economic point of view among the fluids analyzed [21].

Advanced exergetic analyses have focused on the ORC cycle, taking into account the advantage of adapting this cycle to another thermal system for different applications, such as waste heat recovery [22], thermodynamic optimization [23], and emergy analysis [24]. Also, several works have combined these studies to obtain improvement potentials. In applications in turbocharged combustion engines, conventional exergetic analysis gives the evaporator and the expander priority improvement potential while advanced exergy analysis suggests the expander and pump as a priority, and the cycle exergy destruction can be reduced by 36.5% [25]. For applications of advanced exergo-economic analysis taking into account waste heat recovery in geothermal applications, low-temperature solar applications, and waste heat recovery from engine gases, the exergetic efficiency of the ORC improves by 20%, optimizing the system through advanced exergetic analysis and proposing the expander, evaporator, condenser, and pump as improvement potentials. Different organic fluids have been tested in the ORCs to improve their performance, obtaining that pentane, cyclohexane, iso-butene, iso-pentane, and cyclohexane have the highest avoidable endogenous cost corresponding to the heat sources evaluated. In addition, the avoidable endogenous cost is sensitive to the heat source temperature, and it is possible to reduce the heat source temperature increase from 100 to 150 °C by 28% [26]. Therefore, it has been identified that advanced exergetic and thermo-economic analysis is one of the alternatives to achieve technically and economically favorable operating conditions, and to achieve its application in real conditions.

In response to the inadequate management of energy resources in industrial processes, there is a need to improve the efficiency of equipment and processes, in addition to reducing the environmental impact. Thus, the energy recovery of the exhaust line of the natural gas generation engines is one of the alternatives to increase the thermal efficiency of these systems [27]. However, this issue has been approached from different approaches but not articulated with alternative generation systems, which leads to an enormous scientific impact since if it is true that different ORC configurations have been studied, these have not been studied from an advanced exergetic point of view and integrated with thermo-economic modeling in real contexts of operation of stationary high-power natural gas turbocharged engines as a means of heat recovery, in order to obtain technically and economically viable solutions that allow their commercial application [28].

Thus, the main contribution of this work was to perform an advanced thermo-economic analysis of an organic Rankine cycle for a bottoming natural gas engine, and its respective comparison with the results obtained with conventional exergetic and exergo-economic analyses. The analysis of the irreversibilities of each component is presented, and the possible improvements to the cycle are found using the concepts of endogenous/exogenous and avoidable/unavoidable exergy destruction, combinined with the exergo-economic analysis, thus finding the advance cost rate improvement opportunities for each component based on the irreversibilities of the thermal system.

2. Methodology

2.1. Description of the Cycle

The cycle to be analyzed can be seen in Figure 1. The natural gas generation engine operates with an air/natural gas mixture, which is compressed before it enters the cylinders to improve the engine's thermal performance. The exhaust gases are expanded by means of a turbo compressor flow S1 (708 K, 102 kPa), where energy is transferred by means of the heat exchanger 1 (HX1) to the thermal oil in stream S5, and discharged to the environment in stream S2. The thermal oil circulates through the energy supplied by the thermal oil pump (P1), and the hot fluid coming out of HX1 in stream S3 (616 K, 101.4 kPa) operates as a thermal source to evaporate and reheat through the heat exchanger 2 (HX2) the organic fluid, which is toluene in this case study. The maximum values of thee pressure and temperature of the organic Rankine cycle are presented in the turbine inlet (546 K, 675 kPa), where the organic fluid then expands into the S7 stream (475 K, 22 kPa) in the turbine (T1), generating additional energy without increasing the fuel consumption. To complete the thermal cycle, the organic fluid decreases the pressure to its lowest point, passing to the condensation stage from S7 to S8 (338 K, 675 kPa).

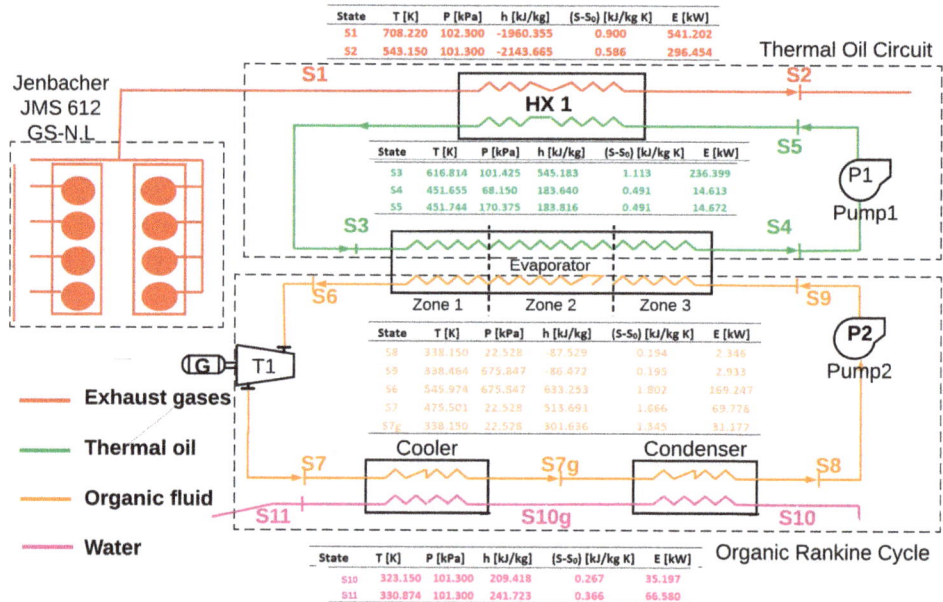

Figure 1. The organic Rankine cycle waste heat recovery system.

The 2 MW Jenbacher engine JMS 612 GS-N. L was modeled and studied, as shown in Figure 2, with its technical specifications and nominal operating conditions [29]. This engine operates with natural gas as fuel, since its high robustness allows it to better adapt to variable load regimes. This engine is widely used for self-generation purposes worldwide and is installed in a company of the plastic sector in the city of Barranquilla, Colombia without any waste heat recovery system. The engine regulates fuel consumption to operate between a minimum load of 1000 kWe and a maximum load of 1982 kWe, with an excess air number (lambda) of 1.79 and 1.97, respectively, generating unused exhaust gases in each of its 12 cylinders with a temperature ranging from 580 to 650 °C.

Figure 2. Jenbacher JMS-612 GS-N.L technical specification.

2.2. Energy and Exergy Analyses

The exergy analysis is defined from the second law of thermodynamics, which means that, unlike the analysis of energy, it depends on the ambient temperature and pressure in which the process in study operates, which allows any system to be investigated in changing environmental conditions. The following assumptions were considered to develop thermodynamic modeling of the RORC:

- The thermal process and component subsystems were assumed as a steady state condition.
- All thermal devices were assumed in adiabatic conditions.
- The pressure drops in the waste heat recovery based on ORC devices and pipelines were neglected.
- The reference temperature for the physical and chemical exergy calculations was 288 K.

The global equation of exergy balance, valid for any volume control system, is shown in Equation (1) [21]:

$$\dot{X}_{heat} + \sum_{i=1}^{n} (\dot{m}_i \cdot \varepsilon_i)_{IN} = P + \sum_{i=1}^{k} (\dot{m}_i \cdot \varepsilon_i)_{OUT} + \dot{E}_{ex,\,D}, \qquad (1)$$

where $\dot{X}_{heat}$ is the exergy of heat transfer in kW, $\dot{m}$ is the mass flow in kg/s, ε is the specific entropy in $\frac{kJ \cdot K}{kg}$, P is the power in kW, and $\dot{E}_{ex,\,D}$ is the exergy destruction [30]. The exergy by heat transfer at temperature T is defined according to Equation (2) [31]:

$$\dot{X}_{heat} = \sum \left(1 - \frac{T_0}{T}\right) \cdot \dot{Q}. \qquad (2)$$

The exergy of a fluid flow stream is defined as the energy power of the fluid flow, with the mass flow ratio of the fluid flow and the pressure and temperature of the fluid flow being necessary, as well as knowing the environmental conditions (pressure and temperature) in which the fluid flow operates [32]. Therefore, the exergetic power of the fluid flow stream was calculated according to Equation (3):

$$\dot{E}_{ex,\,i} = \dot{m}_i \cdot \varepsilon_i, \qquad (3)$$

where the specific exergy (ε_i) was calculated according to Equation (4) [33]:

$$\varepsilon_i = (h_i - h_0) - T_0 \cdot (s_i - s_0). \qquad (4)$$

Therefore, the exergy efficiency (η_{ex}) for a thermal system was calculated according to Equation (5), as a function of the exergy output ($\dot{E}_{out}$) and exergy input ($\dot{E}_{in}$) to the system or device:

$$\eta_{ex} = \frac{\dot{E}_{out}}{\dot{E}_{in}}. \qquad (5)$$

Some energy and exergy performance indicators were calculated for the waste heat recovery system based on the ORC [34], cycle thermal cycle efficiency ($\eta_{I,\,c}$), calculated according to Equation (6); the heat recovery efficiency (ε_{hr}) as shown in Equation (7); and the overall energy conversion efficiency $\left(\eta_{I,\,overall}\right)$, given by Equation (8) [20]:

$$\eta_{I,\,C} = \frac{\dot{W}_{net}}{\dot{Q}_G}, \tag{6}$$

$$\varepsilon_{hr} = \frac{\dot{Q}_G}{\dot{m}_{10} C_{P10}(T_{10} - T_0)}, \tag{7}$$

$$\eta_{I,\,overall} = \eta_{I,\,C} \cdot \varepsilon_{hr}. \tag{8}$$

In addition, to measure the thermal efficiency improvement, the increase in thermal efficiency was calculated through Equation (9), as a function of the net power generated by the ORC ($\dot{W}_{net}$), and the heat supplied by the fuel mass rate ($\dot{m}_{fuel}$):

$$\Delta\eta_{thermal} = \frac{\dot{W}_{net}}{\dot{m}_{fuel} \cdot LHV}. \tag{9}$$

The specific fuel consumption (BSFC) was calculated by the mean of Equation (10) [20], and the absolute reduction in the specific fuel consumption as a consequence of the waste heat recovery was calculated as presented in Equation (11):

$$BSFC_{ORC-engine} = \frac{\dot{m}_{fuel}}{\dot{W}_{engine} + \dot{W}_{net}}, \tag{10}$$

$$\Delta BSFC = \frac{|BSFC_{ORC-engine} - BSFC_{engine}|}{BSFC_{engine}} \cdot 100. \tag{11}$$

2.3. Advanced Exergetic Analysis

In advanced exergetic analysis, the values of exergy destruction are divided into four basic parts: Endogenous, exogenous, avoidable, and unavoidable exergy destruction. Avoidable and unavoidable exergy destruction refers to the system improvement potentials. The destruction of avoidable exergy, $\dot{E}_{D,c}^{AV}$, represents the potential for improvement, during the destruction of unavoidable exergy, $\dot{E}_{D,c}^{UN}$, which represents the limitations. The avoidable part of exergy destruction is described in Equation (12):

$$\dot{E}_{D,c}^{AV} = \dot{E}_{D,c} - \dot{E}_{D,c}^{UN}, \tag{12}$$

where the destruction of unavoidable exergy can be calculated with Equation (13):

$$\dot{E}_{D,c}^{UN} = \dot{E}_{P,c} \left(\frac{\dot{E}_{D,c}}{\dot{E}_{P,c}}\right)^{UN}. \tag{13}$$

The destruction of endogenous exergy, $\dot{E}_{D,c}^{EN}$, and exogenous $\dot{E}_{D,c}^{EX}$ are related to the operational relation between the components of the system. The endogenous part of the exergy destruction is associated only with the irreversibilities that occur in component c, where all the other components operate ideally, and component c operates with its real conditions. On the other hand, the exogenous part of the exergy destruction is produced by the other components. This part can be determined by subtracting the endogenous exergy destruction from the real exergy destruction of component c, as shown in Equation (14):

$$\dot{E}_{D,c}^{EX} = \dot{E}_{D,c} - \dot{E}_{D,c}^{EN}. \tag{14}$$

In addition, the destruction of unavoidable endogenous exergy, $\dot{E}_{D,c}^{UN, EN}$, was calculated by the Equation (15), the destruction of unavoidable exogenous exergy, $\dot{E}_{D,c}^{UN,EX}$, through Equation (16), the destruction of avoidable endogenous exergy, $\dot{E}_{D,c}^{AV, EN}$, with Equation (17), and the destruction of avoidable exogenous exergy, $\dot{E}_{D,c}^{AV,EX}$, by mean of Equation (18) [7]:

$$\dot{E}_{D,c}^{UN, EN} = \dot{E}_{D,c}^{EN} \left(\frac{\dot{E}_{D,c}}{\dot{E}_{P,c}} \right)^{UN}, \tag{15}$$

$$\dot{E}_{D,c}^{UN, EX} = \dot{E}_{D,c}^{UN} - \dot{E}_{D,c}^{UN,EN}, \tag{16}$$

$$\dot{E}_{D,c}^{AV, EN} = \dot{E}_{D,c}^{EN} - \dot{E}_{D,c}^{UN,EN}, \tag{17}$$

$$\dot{E}_{D,c}^{AV, EX} = \dot{E}_{D,c}^{AV} - \dot{E}_{D,c}^{AV,EN}. \tag{18}$$

2.4. Conventional Exergo-Economic Analysis

Exergetic analyses are used to determine the location, type, and magnitude of thermodynamic inefficiencies in system components. On the other hand, exergo-economic analyses combine the concept of exergy and economic analyses to obtain a tool for the optimization of energy systems [4]. In addition, the economic model takes into account the components' cost, including amortization, maintenance, and fuel costs. To define a cost function that depends on interest optimization parameters, the cost of components must be expressed in terms of thermodynamic design parameters. The cost balance equations applied to component c of the system under study show that the sum of rates associated with all outgoing exergy flows equals the sum of cost rates of all incoming exergy flows, plus those corresponding to charges due to capital investment and operating and maintenance costs, as shown in Equation (19) [12]:

$$\sum_e \dot{C}_{e,c} + \dot{C}_{w,c} = \dot{C}_{q,c} + \sum_i \dot{C}_{i,c} + \dot{Z}_c, \tag{19}$$

where the cost rates of input exergy flow ($\dot{C}_i$) are defined in Equation (20), the cost rates of output exergy flow in Equation (21), the heat transfer cost rate in Equation (22), and the cost rate related to energy transfer by work in Equation (23) [35,36]:

$$\dot{C}_i = c_i \cdot \dot{E}_i = c_i [\dot{m}_i e_i], \tag{20}$$

$$\dot{C}_e = c_e \cdot \dot{E}_e = c_e [\dot{m}_e e_e], \tag{21}$$

$$\dot{C}_q = c_q \cdot \dot{E}_q, \tag{22}$$

$$\dot{C}_w = c_w \cdot \dot{W}, \tag{23}$$

where c_i, c_e, c_q, and c_w are the costs per unit of exergy in \$/GJ, and $\dot{Z}_c$ is the sum of the cost rates associated with the cost of capital investment, $\dot{Z}_c^{CI}$, and operation and maintenance costs, $\dot{Z}_c^{OM}$, as shown in Equation (24) [37]:

$$\dot{Z}_c = \dot{Z}_c^{CI} + \dot{Z}_c^{CI} = CRF \cdot \left[\frac{\varphi_r}{N} \cdot 3600 \right] \cdot PEC_c, \tag{24}$$

where the PEC_c is the purchase equipment cost of component c, which is given for all components of the system; N is the number of annual hours that the unit operates; and φ_r is the maintenance factor, which is generally approximately 1.06 [38]. The modeling and sizing of the plate heat exchanger equipment was done for each of the evaporator, condenser, and recovery zones [39]. In addition, thermoeconomic modeling and cost balances for the integrated configuration with the engine were developed by the authors [40]. Also, the CRF is the capital recovery factor, which depends on the interest rate and the estimated lifetime of the equipment, which was calculated based on Equation (25):

$$CRF = \frac{i[1+i]^n}{[1+i]^n - 1}, \qquad (25)$$

where i is the interest rate, and n is the total period of operation of the system in years.

2.5. Advanced Exergo-Economic Análisis

2.5.1. Unavoidable and Avoidable Costs

The avoidable and unavoidable cost ratios associated with the exergy destruction within each component of the system were calculated by Equations (26) and (27), respectively:

$$\dot{C}_{D,c}^{UN} = c_{F,c} \cdot \dot{E}_{D,c}^{UN}, \qquad (26)$$

$$\dot{C}_{D,c}^{AV} = c_{F,c} \cdot \dot{E}_{D,c}^{AV}, \qquad (27)$$

where the sum of the avoidable and unavoidable costs of exergy destruction is equal to the total cost associated with exergy destruction, as shown in Equation (28) [5]:

$$\dot{C}_{D,c} = c_{F,c} \cdot \dot{E}_{D,c} = \dot{C}_{D,c}^{UN} + \dot{C}_{D,c}^{AV}. \qquad (28)$$

The unavoidable investment cost was calculated considering an extremely inefficient version of component c [41]. Therefore, for the calculation of the unavoidable investment cost rate in the components [42], some operational conditions were proposed, as shown in Table 1.

Table 1. Main assumptions for real conditions, ideal conditions, unavoidable exergy, and unavoidable investment cost.

Component	Real Conditions	Theorical Conditions	Unavoidable Exergy Destruction	Unavoidable Investment Costs
Pump 1	$\eta_{iso} = 75\%$	$\eta_{iso} = 100\%$	$\eta_{iso} = 95\%$	$\eta_{iso} = 60\%$
Turbine	$\eta_{iso} = 80\%$	$\eta_{iso} = 100\%$	$\eta_{iso} = 95\%$	$\eta_{iso} = 70\%$
Condenser	$\Delta T_{min} = 15\ °C$	$\eta_{iso} = 70\%$	$\Delta T_{min} = 3\ °C$	$\Delta T_{min} = 26\ °C$
Evaporator	$\Delta T_{min} = 35\ °C$	$\Delta T_{min} = 0\ °C$	$\Delta T_{min} = 3\ °C$	$\Delta T_{min} = 50\ °C$
Pump 2	$\eta_{iso} = 75\%$	$\eta_{iso} = 100\%$	$\eta_{iso} = 95\%$	$\eta_{iso} = 60\%$

The avoidable investment cost was calculated by subtracting the unavoidable cost rate from the total investment cost, as shown in Equation (29):

$$\dot{Z}_c^{AV} = \dot{Z}_c - \dot{Z}_c^{UN}. \qquad (29)$$

2.5.2. Endogenous and Exogenous Cost Rates

The endogenous cost rates $\left(\dot{C}_{D,c}^{EN}\right)$ and exogenous $\left(\dot{C}_{D,c}^{EX}\right)$ are defined according to Equations (30) and (31), respectively:

$$\dot{C}_{D,c}^{EN} = c_{F,c} \cdot \dot{E}_{D,c}^{EN}, \tag{30}$$

$$\dot{C}_{D,c}^{EX} = c_{F,c} \cdot \dot{E}_{D,c}^{EX}, \tag{31}$$

where the sums of the endogenous and exogenous cost rates of exergy destruction are equal to the total cost rate associated with exergy destruction, as defined in Equation (32):

$$\dot{C}_{D,c} = c_{F,c} \cdot \dot{E}_{D,c} = \dot{C}_{D,c}^{EN} + \dot{C}_{D,c}^{EX}. \tag{32}$$

Therefore, the exogenous investment rate was calculated with Equation (33) as follows:

$$\dot{Z}_c^{AV} = \dot{Z}_c - \dot{Z}_c^{UN}. \tag{33}$$

2.5.3. Splitting Cost Rates

The cost rates with respect to exergy desegregation can be calculated with Equations (34)–(37) [6]:

$$\dot{C}_{D,c}^{EN,AV} = c_{F,c} \cdot \dot{E}_{D,c}^{EN,AV}, \tag{34}$$

$$\dot{C}_{D,c}^{EN,UN} = c_{F,c} \cdot \dot{E}_{D,c}^{EN,UN}, \tag{35}$$

$$\dot{C}_{D,c}^{EX,AV} = c_{F,c} \cdot \dot{E}_{D,c}^{EX,AV}, \tag{36}$$

$$\dot{C}_{D,c}^{EX,UN} = c_{F,c} \cdot \dot{E}_{D,c}^{EX,UN}, \tag{37}$$

where $\dot{C}_{D,c}^{EN,AV}$ represents the unavoidable cost rate without component c, associated with the operation of the same component, and the value can be reduced by optimizing the component through technological improvements. Also, the cost $\dot{C}_{D,c}^{EX,AV}$ is the avoidable exogenous cost rate that can be reduced by optimizing other components of the cycle while the costs $\dot{C}_{D,c}^{EN,UN}$ and $\dot{C}_{D,c}^{EX,UN}$ are the unavoidable endogenous and unavoidable exogenous cost rates, respectively.

The endogenous cost rate of component c can be calculated with Equation (38), and the rate of unavoidable endogenous investment costs can be calculated using Equation (39) [7]:

$$\dot{Z}_c^{EN} = \dot{E}_{P,c}^{EN} \left(\frac{\dot{Z}}{\dot{E}_P}\right)_c^{UN}, \tag{38}$$

$$\dot{Z}_c^{EN,UN} = \dot{E}_{P,c}^{EN} \left[\frac{\dot{Z}_c}{\dot{E}_{P,c}}\right]. \tag{39}$$

Therefore, the equations used to calculate the desegregated investment costs are presented from Equations (40) to (42):

$$\dot{Z}_c^{EN,AV} = \dot{Z}_c^{EN} - \dot{Z}_c^{EN,UN}, \tag{40}$$

$$\dot{Z}_c^{EX,UN} = \dot{Z}_c^{UN} - \dot{Z}_c^{EN,UN}, \tag{41}$$

$$\dot{Z}_c^{EX,AV} = \dot{Z}_c^{EX} - \dot{Z}_c^{EN,UN}. \tag{42}$$

3. Results and Discussion

In this section, the influence of the engine load on the heat recovery system energy and exergy performance integrated into the natural gas engine was studied, as an alterntive to reduce the global operational cost and increase the thermal efficiency [43]. The engine power control system adjusts internal engine variables, such as the pressure and temperature of the air-fuel mixture before entering the cylinders, and the recirculation percentage, to provide high efficiency in partial load operation of the gas engine. Some energy indicators were proposed to study the performance of the waste heat recovery system based on ORC, as shown in Figure 3, while the evaporating pressure was set to 675.8 kPa, and toluene was selected as the working fluid [33]. For safety restriction, all feasible operating points of the proposed system at different engine loads ensured that toluene evaporates completely at the outlet of the evaporator to prevent corrosion of the liquid in the expander, in addition to a gas temperature at the outlet of the evaporator (state 11) being higher than the acid dew temperature (200 °C) to avoid acidic corrosion of the exhaust [34].

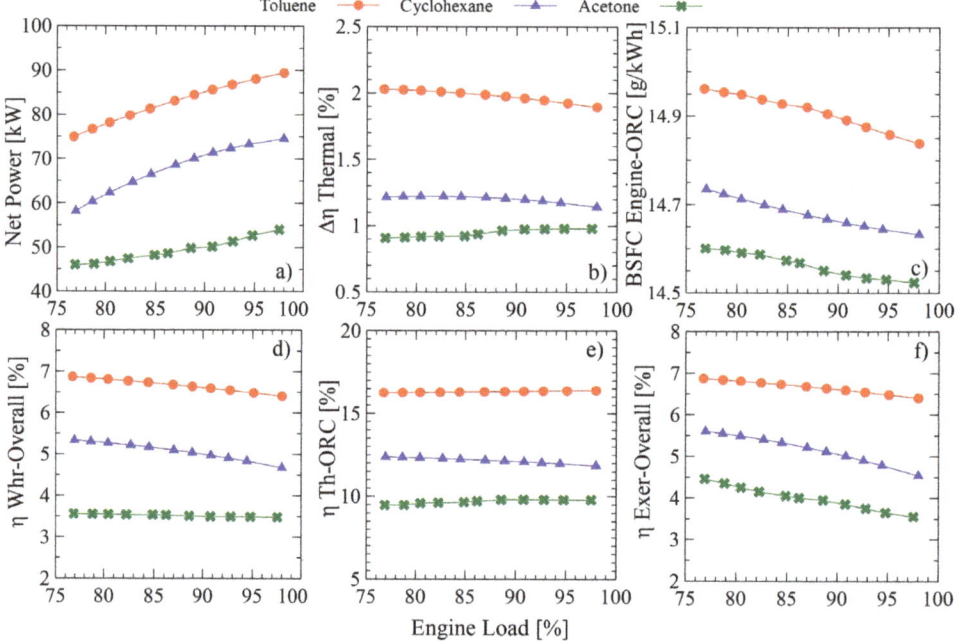

Figure 3. Energy and exergy indicator of the waste heat recovery system with different organic fluids and engine load, (**a**) met power, (**b**) Absolute increase in thermal efficiency, (**c**) specific fuel consumption, (**d**) global energy conversion efficiency, (**e**) ORC thermal efficiency, and (**f**) global exergetic efficiency.

The results show that the absolute increase in thermal efficiency (Figure 3b) decreases for toluene and cyclohexane, as does the overall energy conversion efficiency (Figure 3d) with an increasing engine load, while the net power output (Figure 3a) presents its maximum value with toluene (89.4 kW—97.9%), cyclohexane (73.2 kW—97.9%), and acetone (53.2 kW—91.81%), respectively. However, in an engine operating interval, acetone presents a slight increase in thermal efficiency, and then decreases, presenting a maximum at an 82.68% engine load.

These results are due to a higher engine load, implying an increase in the exhaust gas flow according to the first and second laws of thermodynamics [44] while a greater energy loss is presented in the recuperator heat exchanger 1 (HX1) because of the evaporation pressure and thermal oil temperature have been limited. As the engine load increases, there is an increase in the fluid evaporating temperature at the evaporator. Therefore, the power increases, which is the main factor for thermal and exergetic efficiency. However, the isentropic turbine efficiency decreases slightly as a consequence of the increase in the thermal oil temperature, causing a decrease in the energy indicators at high engine loads. Likewise, the tendency to increase the power with the engine load is a consequence of both the increase in the inlet thermal oil temperature to the evaporator, which leads to an increase in the toluene mass flow, and the enthalpy difference between the outlet and the inlet of the pump and turbine, but this is more relevant in the turbine.

In addition, the results obtained from the traditional exergetic and exergo-economic analysis are shown in Table 2, where the exergy and fraction of exergy destroyed, $y_{D,c}$, shows that the greatest values are present in the heat exchanger 1 (shell and tube heat exchanger) with 32.54%, the evaporator (28.32%), and the condenser with 27.97%. The component with the highest destroyed exergy value (41.95 kW) is heat exchanger 1, being one of the components with the lowest exergetic efficiency of the cycle, due to the large heat exchanger area required and the high temperature difference. The greater the investment and the cost of exergy destroyed, the greater the influence of the component in the system, therefore, the component with the greatest improvement in cost efficiency of the total plant can be defined. In the case study, the components with the greatest opportunities for improvement in this ratio are the condenser and HX 1. Therefore, these components are the most important from a thermodynamic point of view.

Table 2. The results of conventional analysis for all components in the waste heat recovery system.

Components	E_f [kW]	E_p [kW]	E_d [kW]	E_{loss} [kW]	E [%]	$Y_{d,k}$	C_f [USD/GJ]	C_p [USD/GJ]	C_d [USD/h]	C_{loss} [USD/h]	Z [USD/h]	$Z + C_d + C_{loss}$ [USD/h]	f_c [%]
HX 1	541.20	202.79	41.95	338.40	37.47	32.54	15.22	11.97	2.30	16.24	2.67	21.22	53.79
P1	0.37	0.05	0.31	-	15.60	0.24	47.56	1801.58	0.05	-	0.29	0.34	85.24
Turb	99.48	85.58	13.89	-	86.03	10.77	19.11	47.85	0.95	-	7.89	8.85	89.20
P2	0.75	0.58	0.16	-	77.60	0.13	47.85	197.65	0.02	-	0.28	0.31	90.78
Evap	202.85	166.34	36.51	-	82.00	28.32	12.46	18.48	1.63	-	1.96	2.60	54.58
Cond	-	-	36.05	66.58	-	27.97	55.48	19.11	7.18	13.27	1.61	22.07	18.35

The exergo-economic factor, f_c, is the effective parameter that allows us to compare and evaluate the components that make up the system. A high value for this parameter indicates that for the component under study, acquisition costs predominate over operation and maintenance costs. For example, in the case of the condenser, which is the component with the lowest value of the exergo-economic factor, it can be concluded that expenses are mostly related to operating and maintenance costs compared to acquisition costs.

By means of the advanced exergetic analysis, the exergy destruction can be disaggregated for each one of the components. In this way, the real possibilities of improvement can be determined both through the operational and design point of view of the component, and the global consideration of the thermal system. From the solution of Equations (12)–(18), and the unavoidable operation conditions described in Table 1, the disaggregation of the exergy can be found in its endogenous, exogenous, avoidable, and unavoidable part, as well as avoidable and unavoidable endogenous and its avoidable and unavoidable exogenous counterpart. The determination of the avoidable part of the destroyed exergy is a significant step because it allows identification of opportunities for improvement in the component and its interaction with the rest of the components. Also, this result allows knowledge of which is the optimal way to increase the thermal efficiency of the system, besides providing valuable information about how the components operate together as a global system. Figure 4 presents a graphical version of the improvement opportunities in each component from the exergetic point.

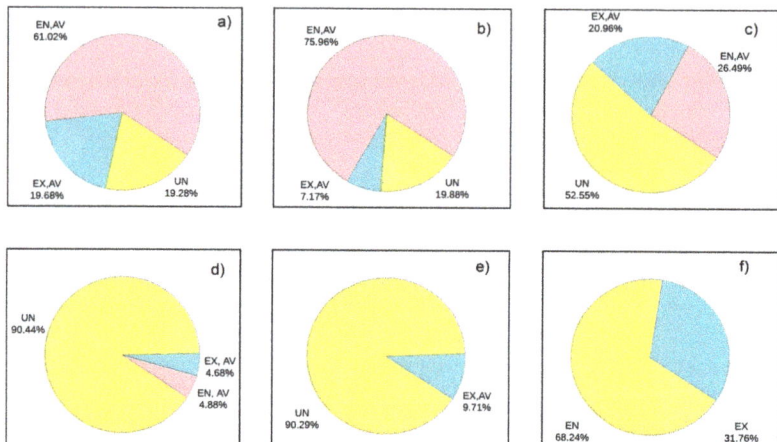

Figure 4. Advanced exergy analyses for each component in the ORC cycle, (**a**) HX 1, (**b**) P1, (**c**) turbine, (**d**) P2, (**e**) evaporator, and (**f**) condenser.

The results of the advanced exergetic analysis and economic exergetic analysis are presented in Table 3, where the disaggregation of the destroyed exergy was calculated as a function of the endogenous, exogenous, avoidable, and unavoidable for each of the components under study. The results show that most of the destroyed exergy is endogenous (78.53% of the total destroyed exergy), emphasizing that the interaction between components does not have a significant effect on the overall exergetic performance of the cycle. Similarly, it is noted that the component with the greatest avoidable exergy destroyed in the system is the turbine, with a value of 11.075 kW, where 69.625% is endogenous and 30.374 is exogenous, which means that in the turbine, there is a real great opportunity for improvement. On the other hand, in the unavoidable part, the components with the greatest technological limitations are the HX 1 and evaporator, representing 96.1% of the total unavoidable exergy of the cycle.

Table 3. Splitting exergy destruction for each component.

Components	$E_{D,c}^{EN}$ [kW]	$E_{D,c}^{EX}$ [kW]	$E_{D,c}^{AV}$ [kW]	$E_{D,c}^{UN}$ [kW]	$E_{D,c}^{UN,EN}$ [kW]	$E_{D,c}^{UN,EX}$ [kW]	$E_{D,c}^{AV,EN}$ [kW]	$E_{D,c}^{AV,EX}$ [kW]
HX 1	37.339	4.615	4.000	37.953	0.000	0.000	0.000	0.000
Pump 1	0.159	0.157	0.011	0.030	0.015	0.290	0.144	−0.133
Turbine	8.661	5.229	11.075	2.819	0.095	1.869	7.711	3.360
Pump 2	0.140	0.029	0.140	0.028	0.011	0.016	0.013	0.012
Evaporator	26.849	9.663	3.542	32.971	0.000	0.000	0.000	0.000
Condenser	24.607	11.451	0.000	0.000	0.000	0.000	0.000	0.000
Total	97.755	31.144	18.768	73.801	0.121	2.175	7.868	3.239
%	75.83%	24.16%	-	-	0.14%	6.69%	8.14%	10.79%

The equations presented in Sections 2.5.2 and 2.5.3 were used to calculate the advance exergy destruction costs as shown in Table 4, which is based on the result of the advanced destroyed exergy.

It can be observed that the endogenous exergy destruction is higher than the exogenous cost in the components of the thermal cycle, which is a consequence of the high endogenous investment costs values for all components of the system with respect to the exogenous investment cost, as shown in Table 5. Therefore, it can be established that the interaction between components in terms of investment costs is not very relevant in the system; however, for the component under study, it is a parameter of vital importance. Also, it can be observed that the rates of unavoidable investment costs for the components studied showed an inclination in the unavoidable part.

Table 4. Advanced exergy destruction cost rates for all components in the waste heat recovery system.

Components	$C_{D,k}$ [USD/h]	$C_{D,k}^{EN}$ [USD/h]	$C_{D,k}^{EX}$ [USD/h]	$C_{D,k}^{AV}$ [USD/h]	$C_{D,k}^{UN}$ [USD/h]	$C_{D,k}^{AV,EN}$ [USD/h]	$C_{D,k}^{AV,EX}$ [USD/h]	$C_{D,k}^{UN,EN}$ [USD/h]	$C_{D,k}^{UN,EX}$ [USD/h]
HX 1	2.300	2.047	0.253	0.219	2.080	0.000	0.000	0.000	0.000
Pump 1	0.051	0.027	0.027	0.005	0.049	0.025	−0.019	0.002	0.047
Turbine	0.956	0.591	0.3599	0.701	0.254	0.531	0.171	0.065	0.188
Pump 2	0.029	0.024	0.005	0.024	0.005	0.022	0.002	0.002	0.003
Evaporator	1.638	1.205	0.433	0.159	1.479	0.000	0.000	0.000	0.000
Condenser	7.188	4.906	2.283	0.000	0.000	0.000	0.000	0.000	0.000

Table 5. Advanced investment costs for all components in the WHR (Wast Heat Recovery).

Components	$Z_{d,k}$ [USD/h]	Z^{EN} [USD/h]	Z^{EX} [USD/h]	Z^{AV} [USD/h]	Z^{UN} [USD/h]	$Z^{AV,EN}$ [USD/h]	$Z^{AV,EX}$ [USD/h]	$Z^{UN,EN}$ [USD/h]	$Z^{UN,EX}$ [USD/h]
HX 1	2.678	2.648	0.030	−0.046	2.724	0.251	−0.297	2.397	0.328
Pump 1	0.295	0.288	0.007	0.001	0.294	−0.170	0.171	0.458	−0.164
Turbine	7.898	5.417	2.480	2.215	5.683	0.053	2.161	5.364	0.319
Evaporator	1.970	1.957	0.013	0.022	1.948	0.025	−0.004	1.932	0.016
Condenser	1.760	1.645	0.115	0.100	1.660	-	-	-	-

Negative values of exogenous investment cost rates (Z^{EX}, $Z^{AV,EX}$, $Z^{UN,EX}$) revealed that investment costs within these components might decrease if investment costs within the other components are increased.

In order to show a comparison of the destroyed exergy relationship between traditional and advanced exergetic analysis, Figure 5 is shown. In Figure 5A, a slight difference of the parameter under study is denoted because the one that was calculated by means of the advanced exergetic analysis only emphasizes the exergy that is destroyed by each component, that is to say, the endogenous part (without the interaction that this one has with its surroundings).

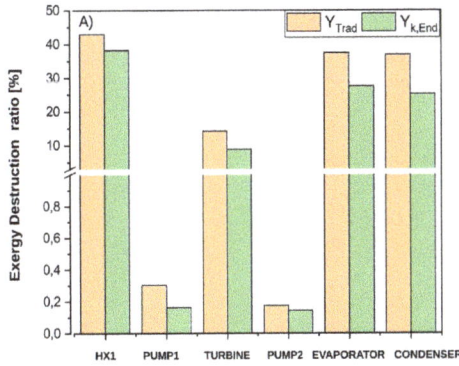
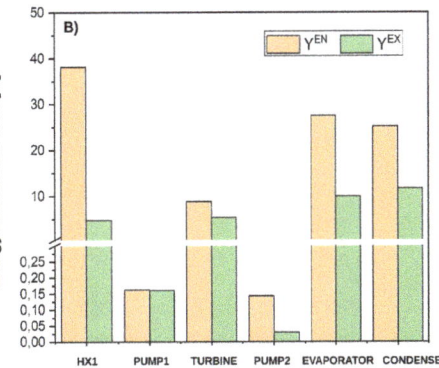

Figure 5. Contribution of each component to the overall exergy destruction of the cycle based on (**A**) traditional and (**B**) advanced exergy analysis.

However, the relationship of exergy destroyed by the component was also calculated by emphasizing which fraction is borne by the component itself or by the interaction of the component with its surroundings, as shown in Figure 5B. From this graph, what has been mentioned before is supported, that is, that the interaction between each of the components of the system is not significant in comparison to the exergy that destroys the component under its own operating conditions. A comparative analysis was performed by implementing a new exergo-economic factor calculated by advanced exergetic analysis, as shown in Figure 6.

As a percentage, it can be seen that the exergo-economic factor, as well as the traditional and advanced approach, presents a similarity in the components that make up the system. So, the main efforts should concentrate on designing the most efficient heat exchangers [35], with a smaller heat transfer area and less exergy destruction, without increasing the purchase equipment costs.

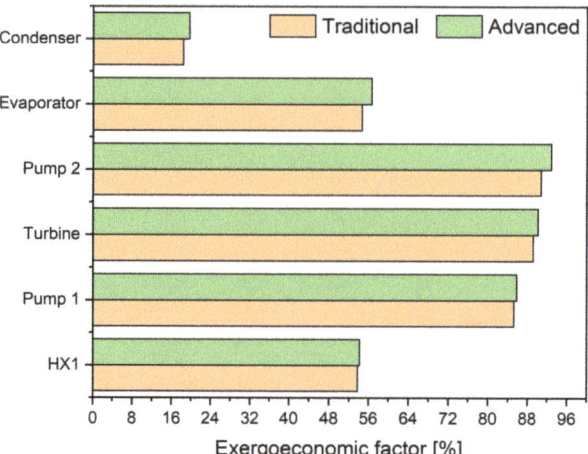

Figure 6. Comparison of the calculated exergo-economic factor based on traditional and advanced exergy analysis.

4. Conclusions

In this paper, the benefit offered by developed traditional and advanced exergetic analysis in thermal systems was shown, in particular in the organic Rankine cycle systems. Exergetic analysis allows determination of the sources of irreversibility in a thermal system, and therefore indicates the starting points of an optimization procedure, and contributes to the rational use of the energetic resources. In the study carried out, it was possible to determine which equipment that resulted in greater destruction of exergy introduced in the waste heat recovery system based on the organic Rankine cycle. The equipment in which the design or operational improvements can be made was also determined, since the implementation of some recommendations is not practical for optimizing the cycle due to operational or design limitations. Therefore, traditional exergy, advanced exergy, and exergo-economic analysis were applied to gain a better understanding of the system performance. Moreover, a comprehensive comparison was conducted to further assess the system from various points of view.

The conventional exergy showed that the heat exchanger 1 had the largest exergy destruction and exergy destruction, and highest investment costs (41.95 kW, 32.54%, and 2.67 USD/h). The results of the energetic and exergetic analysis of the system showed that the exergy destroyed is a measure of the degree of process irreversibility. Thus, in the case of heat exchanger 1, the causes of the irreversibility were due to the heat transfer through a finite temperature difference higher than 100 °C. Similarly, the results of exergy destructions appeared to be in accordance with the exergy efficiencies. That is, a smaller exergy efficiency implies greater exergy destruction in the system components.

Also, the highest exergo-economic factor was found in the pump 2, turbine, and pump 1, with 90.78%, 89.20%, and 85.24%, respectively. These results were a consequence of the high effect of the purchased equipment cost, and the low thermodynamic efficiency in the aforementioned devices, where the probable solution could be the implementation of low-cost components, which are usually characterized by a lower energy efficiency.

Most of the exergy destruction calculated was endogenous (78.53%), emphasizing that the interaction between components does not have a significant effect on the overall exergetic performance of the cycle. The maximum unavoidable exergy was found for the heat exchanger 1, with 90.44%.

This indicates that there are not too many ways to improve this component. Nevertheless, other components, such as pump 2, pump 1, and turbine 1, have the minimum unavoidable exergy destruction, with 0.028, 0.03, and 2.819 kW. In addition, the component with the highest cost rate was the condenser with 7.188 USD/h, followed by the heat exchanger 1 with 2.3 USD/h, but the highest avoidable cost rate was found for the turbine with a value of 0.701 USD/h.

On the other hand, the advanced exergo-economic analyses showed that the turbine is the component with the major purchase equipment cost in the system, with a value of 7.898 USD/h, which is 54.09% of the total equipment cost of the system. For all components studied, the endogenous investment cost was higher than the exogenous part, showing the weak relation between them. A comparison was realized between the traditional and advanced exergo-economic factor, which resulted in a similar effect in each component, but the advanced exergy approach presented a slightly higher value, implying that the advanced exergetic analysis gives greater precision in terms of results without ignoring the really great opportunities for improvement.

Author Contributions: Conceptualization: G.V.O.; Methodology: G.V.O. and J.D.F.; Software: G.V.O., J.P.R. and J.D.F.; Validation: G.V.O., J.P.R. and J.D.F.; Formal Analysis: G.V.O., J.P.R. and J.D.F.; Investigation: G.V.O.; Resources: G.V.O.; Writing—Original Draft Preparation: G.V.O. and J.P.R.; Writing—Review and Editing: G.V.O. and J.D.F.; Funding Acquisition: G.V.O. All authors have read and agreed to the published version of the manuscript.

Funding: This research received no external funding.

Acknowledgments: Acknowledgments to Universidad del Atlántico, Universidad Francisco de Paula Santander, and the E2 Energía Eficiente S.A E.S. P company by the support received to conduct this research.

Conflicts of Interest: The authors declare no conflict of interest.

Abbreviations

The following abbreviations are used in this manuscript:

ICE	Internal Combustion Engine
HX 1	Heat Exchanger 1
HX2	Heat Exchanger 2
ORC	Organic Rankine Cycle
PP	Pinch Point
WHR	Waste heat recovery
CRF	Capital Recovery Factor
PEC_c	Equipment Purchase Cost of component C

Nomenclature

E	Exergy
h	Enthalpy
$\dot{m}$	Fuel mass rate
P	Pressure
Q	Heat
s	Entropy
Rp	Pressure ratio
T	Temperature
W	Power
ϵ	Exergy efficiency
η	Energy efficiency
y	Exergy destruction ratio
$\dot{Z}_c$	Investment costs
$\dot{C}_{D,c}^{EN}$	Endogenous exergy destruction cost rates
$\dot{C}_{D,c}^{EX}$	Exogenous exergy destruction cost rates
N	Number of annual operation hours

Subscripts

0	References condition
Cond	Condenser
ch	Chemical
D	Destruction
Evap	Evaporator
F	Fuel
iso	Isoentropic
k	Component
min	Minimum
P	Product
ph	Physical
Pump	Pump
Th	Theorical
Tot	Total
Turb	Turbine

Superscripts

AV	Avoidable
EN	Endogenous
EX	Exogenous
EN, AV	Endogenous avoidable
EN, UN	Endogenous unavoidable
EX, AV	Exogenous avoidable
EX, UN	Exogenous unavoidable
id	Ideal
RS	Real
UN	Unavoidable

References

1. Petrakopoulou, F.; Tsatsaronis, G.; Morosuk, T.; Carassai, A. Conventional and advanced exergetic analyses applied to a combined cycle power plant. *Energy* **2012**, *41*, 146–152. [CrossRef]
2. Morosuk, T.; Tsatsaronis, G. Advanced exergy-based methods used to understand and improve energy-conversion systems. *Energy* **2019**, *169*, 238–246. [CrossRef]
3. Anvari, S.; Khoshbakhti Saray, R.; Bahlouli, K. Conventional and advanced exergetic and exergoeconomic analyses applied to a tri-generation cycle for heat, cold and power production. *Energy* **2015**, *91*, 925–939. [CrossRef]
4. Tsatsaronis, G. *Strengths and Limitations of Exergy Analysis BT-Thermodynamic Optimization of Complex Energy Systems*; Bejan, A., Mamut, E., Eds.; Springer: Dordrecht, The Netherlands, 1999; pp. 93–100, ISBN 978-94-011-4685-2.
5. Tsatsaronis, G.; Park, M.-H. On avoidable and unavoidable exergy destructions and investment costs in thermal systems. *Energy Convers. Manag.* **2002**, *43*, 1259–1270. [CrossRef]
6. Kelly, S. Energy Systems Improvement based on Endogenous and Exogenous Exergy Destruction. Ph.D. Thesis, Technische Universität Berlin, Berlin, Germany, 2008.
7. Petrakopoulou, F. Comparative Evaluation of Power Plants with CO_2 Capture: Thermodynamic, Economic and Environmental Performance. Ph.D. Thesis, Technische Universität Berlin, Berlin, Germany, 2010.
8. Long, R.; Bao, Y.J.; Huang, X.M.; Liu, W. Exergy analysis and working fluid selection of organic Rankine cycle for low grade waste heat recovery. *Energy* **2014**, *73*, 475–483. [CrossRef]
9. Long, R.; Kuang, Z.; Li, B.; Liu, Z.; Liu, W. Exergy analysis and performance optimization of Kalina cycle system 11 (KCS-11) for low grade waste heat recovery. *Energy Procedia* **2019**, *158*, 1354–1359. [CrossRef]
10. Tian, H.; Shu, G.; Wei, H.; Liang, X.; Liu, L. Fluids and parameters optimization for the organic Rankine cycles (ORCs) used in exhaust heat recovery of Internal Combustion Engine (ICE). *Energy* **2012**, *47*, 125–136. [CrossRef]

11. Zare, V. A comparative exergoeconomic analysis of different ORC configurations for binary geothermal power plants. *Energy Convers. Manag.* **2015**, *105*, 127–138. [CrossRef]
12. Kerme, E.D.; Orfi, J. Exergy-based thermodynamic analysis of solar driven organic Rankine cycle. *J. Therm. Eng.* **2015**, *1*, 192–202. [CrossRef]
13. Boyano, A.; Morosuk, T.; Blanco-Marigorta, A.M.; Tsatsaronis, G. Conventional and advanced exergoenvironmental analysis of a steam methane reforming reactor for hydrogen production. *J. Clean. Prod.* **2012**, *20*, 152–160. [CrossRef]
14. Morozyuk, T.; Tsatsaronis, G. Strengths and Limitations of Advanced Exergetic Analyses. In Proceedings of the ASME International Mechanical Engineering Congress and Exposition, San Diego, CA, USA, 15–21 November 2013; Volume 6.
15. Mortazavi, A.; Ameri, M. Conventional and advanced exergy analysis of solar flat plate air collectors. *Energy* **2018**, *142*, 277–288. [CrossRef]
16. Mohammadi, Z.; Fallah, M.; Mahmoudi, S.M.S. Advanced exergy analysis of recompression supercritical CO_2 cycle. *Energy* **2019**, *178*, 631–643. [CrossRef]
17. Yue, T.; Lior, N. Exergo-economic competitiveness criteria for hybrid power cycles using multiple heat sources of different temperatures. *Energy* **2017**, *135*, 943–961. [CrossRef]
18. Marami Milani, S.; Khoshbakhti Saray, R.; Najafi, M. Exergo-economic analysis of different power-cycle configurations driven by heat recovery of a gas engine. *Energy Convers. Manag.* **2019**, *186*, 103–119. [CrossRef]
19. Petrakopoulou, F.; Tsatsaronis, G.; Morosuk, T. Advanced Exergoeconomic Analysis of a Power Plant with CO2 Capture. *Energy Procedia* **2015**, *75*, 2253–2260. [CrossRef]
20. Ambriz-Díaz, V.M.; Rubio-Maya, C.; Ruiz-Casanova, E.; Martínez-Patiño, J.; Pastor-Martínez, E. Advanced exergy and exergoeconomic analysis for a polygeneration plant operating in geothermal cascade. *Energy Convers. Manag.* **2019**, 112227. [CrossRef]
21. Nazari, N.; Heidarnejad, P.; Porkhial, S. Multi-objective optimization of a combined steam-organic Rankine cycle based on exergy and exergo-economic analysis for waste heat recovery application. *Energy Convers. Manag.* **2016**, *127*, 366–379. [CrossRef]
22. Mikielewicz, D.; Wajs, J.; Ziółkowski, P.; Mikielewicz, J. Utilisation of waste heat from the power plant by use of the ORC aided with bleed steam and extra source of heat. *Energy* **2016**, *97*, 11–19. [CrossRef]
23. Scaccabarozzi, R.; Tavano, M.; Invernizzi, C.M.; Martelli, E. Thermodynamic Optimization of heat recovery ORCs for heavy duty Internal Combustion Engine: Pure fluids vs. zeotropic mixtures. *Energy Procedia* **2017**, *129*, 168–175. [CrossRef]
24. Zhang, H.; Guan, X.; Ding, Y.; Liu, C. Emergy analysis of Organic Rankine Cycle (ORC) for waste heat power generation. *J. Clean. Prod.* **2018**, *183*, 1207–1215. [CrossRef]
25. Galindo, J.; Ruiz, S.; Dolz, V.; Royo-Pascual, L. Advanced exergy analysis for a bottoming organic rankine cycle coupled to an internal combustion engine. *Energy Convers. Manag.* **2016**, *126*, 217–227. [CrossRef]
26. Dai, B.; Zhu, K.; Wang, Y.; Sun, Z.; Liu, Z. Evaluation of organic Rankine cycle by using hydrocarbons as working fluids: Advanced exergy and advanced exergoeconomic analyses. *Energy Convers. Manag.* **2019**, *197*, 111876. [CrossRef]
27. Valencia, G.; Acevedo, C.; Duarte, J. Thermoeconomic optimization with PSO algotithm of waste heat recovery system based on Organic Rankine Cycle system for a natural engine. *Energies* **2019**, *12*, 4165. [CrossRef]
28. Ochoa, G.V.; Peñaloza, C.A.; Rojas, J.P. Thermoeconomic modelling and parametric study of a simple orc for the recovery of waste heat in a 2 MW gas engine under different working fluids. *Appl. Sci.* **2019**, *9*, 4526. [CrossRef]
29. Ochoa, G.V.; Isaza-Roldan, C.; Forero, J.D. A phenomenological base semi-physical thermodynamic model for the cylinder and exhaust manifold of a natural gas 2-megawatt four-stroke internal combustion engine. *Heliyon* **2019**, *5*, e02700. [CrossRef] [PubMed]
30. Regulagadda, P.; Dincer, I.; Naterer, G.F. Exergy analysis of a thermal power plant with measured boiler and turbine losses. *Appl. Therm. Eng.* **2010**, *30*, 970–976. [CrossRef]
31. Ahmadi, G.; Toghraie, D.; Azimian, A.; Akbari, O.A. Evaluation of synchronous execution of full repowering and solar assisting in a 200MW steam power plant, a case study. *Appl. Therm. Eng.* **2017**, *112*, 111–123. [CrossRef]

32. Taner, T.; Sivrioglu, M. Energy–exergy analysis and optimisation of a model sugar factory in Turkey. *Energy* **2015**, *93*, 641–654. [CrossRef]
33. Ahmadi, G.; Toghraie, D.; Akbari, O.A. Solar parallel feed water heating repowering of a steam power plant: A case study in Iran. *Renew. Sustain. Energy Rev.* **2017**, *77*, 474–485. [CrossRef]
34. Valencia, G.; Fontalvo, A.; Cárdenas, Y.; Duarte, J.; Isaza, C. Energy and exergy analysis of different exhaust waste heat recovery systems for natural gas engine based on ORC. *Energies* **2019**, *12*, 2378. [CrossRef]
35. Quoilin, S.; Aumann, R.; Grill, A.; Schuster, A.; Lemort, V.; Spliethoff, H. Dynamic modeling and optimal control strategy of waste heat recovery Organic Rankine Cycles. *Appl. Energy* **2011**, *88*, 2183–2190. [CrossRef]
36. Sayyaadi, H. Multi-objective approach in thermoenvironomic optimization of a benchmark cogeneration system. *Appl. Energy* **2009**, *86*, 867–879. [CrossRef]
37. Ghaebi, H.; Amidpour, M.; Karimkashi, S.; Rezayan, O. Energy, exergy and thermoeconomic analysis of a combined cooling, heating and power (CCHP) system with gas turbine prime mover. *Int. J. Energy Res.* **2011**, *35*, 697–709. [CrossRef]
38. Bejan, A.; Tsatsaronis, G.; Moran, M.J. *Thermal Design & Optimization*; John Wiley & Sons: Hoboken, NJ, USA, 1995; ISBN 0-471-58467-3.
39. Valencia, G.; Núñez, J.; Duarte, J. Multiobjective optimization of a plate heat exchanger in a waste heat recovery organic rankine cycle system for natural gas engines. *Entropy* **2019**, *21*, 655. [CrossRef]
40. Valencia, G.; Duarte, J.; Isaza-Roldan, C. Thermoeconomic analysis of different exhaust waste-heat recovery systems for natural gas engine based on ORC. *Appl. Sci.* **2019**, *9*, 4017. [CrossRef]
41. Morosuk, T.; Tsatsaronis, G. Advanced Exergoeconomic Analysis of a Refrigeration Machine: Part 1—Methodology and First Evaluation. In Proceedings of the ASME 2011 International Mechanical Engineering Congress and Exposition, Denver, CO, USA, 11–17 November 2011; pp. 47–56.
42. Morosuk, T.; Tsatsaronis, G. Advanced Exergoeconomic Analysis of a Refrigeration Machine: Part 2—Improvement. In Proceedings of the ASME 2011 International Mechanical Engineering Congress and Exposition, Denver, CO, USA, 11–17 November 2011; pp. 57–65.
43. Budes, F.B.; Ochoa, G.V.; Escorcia, Y.C. Hybrid PV and Wind grid-connected renewable energy system to reduce the gas emission and operation cost. *Contemp. Eng. Sci.* **2017**, *10*, 1269–1278. [CrossRef]
44. Diaz, G.A.; Forero, J.D.; Garcia, J.; Rincon, A.; Fontalvo, A.; Bula, A.; Padilla, R.V. Maximum power from fluid flow by applying the first and second laws of thermodynamics. *J. Energy Resour. Technol.* **2017**, *139*, 032903. [CrossRef]

© 2020 by the authors. Licensee MDPI, Basel, Switzerland. This article is an open access article distributed under the terms and conditions of the Creative Commons Attribution (CC BY) license (http://creativecommons.org/licenses/by/4.0/).

Article

A Li-Ion Battery Thermal Management System Combining a Heat Pipe and Thermoelectric Cooler

Chuanwei Zhang [1], Zhan Xia [1,*], Bin Wang [2], Huaibin Gao [1], Shangrui Chen [1], Shouchao Zong [1] and Kunxin Luo [1]

1. School of Mechanical Engineering, Xi'an University of Science and Technology, Xi'an 710054, China; zhangcw@xust.edu.cn (C.Z.); gaohuaibin@xust.edu.cn (H.G.); 17205216074@stu.xust.edu.cn (S.C.); 18205216092@stu.xust.edu.cn (S.Z.); 18205018011@stu.xust.edu.cn (K.L.)
2. College of Engineering, Design and Physical Sciences, Brunel University, London UB8 3PH, UK; Bin.Wang@brunel.ac.uk
* Correspondence: 18205216077@stu.xust.edu.cn

Received: 29 November 2019; Accepted: 11 February 2020; Published: 14 February 2020

Abstract: The temperature of electric vehicle batteries needs to be controlled through a thermal management system to ensure working performance, service life, and safety. In this paper, TAFEL-LAE895 100Ah ternary Li-ion batteries were used, and discharging experiments at different rates were conducted to study the surface temperature increasing characteristics of the battery. To dissipate heat, heat pipes with high thermal conductivity were used to accelerate dissipating heat on the surface of the battery. We found that the heat pipe was sufficient to keep the battery temperature within the desired range with a midlevel discharge rate. For further improvement, an additional thermoelectric cooler was needed for a high discharge rate. Simulations were completed with a battery management system based on a heat pipe and with a combined heat pipe and thermoelectric cooler, and the results were in line with the experimental results. The findings show that the combined system can effectively reduce the surface temperature of a battery within the full range of discharge rates expected in the battery used.

Keywords: thermal management; Li-ion battery; heat pipe; thermoelectric cooler

1. Introduction

With increasing environmental pollution and fossil fuel depletion, electric vehicles are gradually appearing in the automotive market, as they produce no pollution and use electricity instead of fuel [1,2]. As the power source of electric vehicles, Li-ion batteries are characterized by high energy density, low self-discharge rate, no memory effect, and a small size, so they are favored by automobile manufacturers [3–5]. However, electric vehicles are at risk of spontaneous combustion, attributed to the high surface temperature of the Li-ion battery used in electric vehicles, causing thermal runaway where the temperature of the batteries rises constantly, even causing fire. As a result, the battery does not work normally [6]. During the Li-ion battery charging and discharging process, complex chemical reactions occur in its interior and heat is generated, so the temperature of the battery continuously rises [7]. If the heat cannot be dissipated in time, thermal runaway of the Li-ion battery may occur [6]. Therefore, a battery thermal management system (BTMS) to control the temperature of the battery is essential to ensure the normal working and the safety of electric vehicles.

Some studies have been conducted on BTMS, and various cooling methods for Li-ion batteries have been proposed. For an air-cooling system (ACS), where air is used as the cooling medium, the battery pack is primarily cooled by heat convection. However, the ACS is unable to meet the cooling demands of the battery [8]. A liquid cooling system (LCS) for batteries was also developed to cool or heat the battery pack. Rao et al. [9] designed a BTMS based on liquid cooling for a cylindrical Li-ion battery pack and researched the change of in-battery contact surface to determine the length of pack block when the flowing rate of cooling liquid is at the optimal efficiency. Under a battery heat generation of 30 W, Pesaran et al. [10] tested oil and air as cooling media. Their results showed that the surface temperature of the battery in the oil cooling system is nearly 10 °C lower than in ACS under the same working conditions.

Versatile materials have gradually been developed. Researchers explored materials with special characteristics that can be used in BTMS and applied phase change heat transfer media to BTMS. The principle involves using the characteristics of phase change materials (PCMs) to store or release energy through a phase change without significantly changing the temperature. At high discharge rates, Sabbah et al. [11] conducted experiments using 18,650 batteries to demonstrate that the thermal management effect of PCM cooling is better than that of air. Rao and Zhang et al. [12–14] studied different models and concluded that by allowing batteries to be in direct contact with the PCM, the surface temperature of the battery can be reduced by 18 °C. Wu et al. [15] constructed a new type of reinforced composite phase change material (CPCM) based on copper mesh, paraffin, and expanded graphite for BTMS. The heat dissipation and temperature uniformity of the CPCM were better than that of the traditional air convection and PMC.

In addition to the above studies, a heat pipe (HP) with high thermal conductivity in the longitudinal direction and isothermal properties in heat conductive direction, and a thermoelectric cooler (TEC) characterized by refrigeration ability, rapidly transferring heat, small size, and high reliability, were both investigated [16–18]. Joshua et al. [19] applied the HP to BTMS, and found it can better control the temperature of the Li-ion battery than the traditional liquid cooling system. Deng et al. [20] combined an L-style HP with an aluminum plate to build a BTMS and showed that with ambient temperature increasing, the heat dissipation from HP increases and the increasing rate of battery temperature reduces. Wang [18] selected the optimal working current of the TEC and built a BTMS based on it. The results showed the advantage of controlling the temperature of the TEC. High-temperature environments have no significant effect on the performance of the TEC. Lu et al. [21] verified that the TEC is highly effective for heating or cooling the battery pack. Shashank et al. [22] arranged battery cells in an inverted position within a battery pack filled with a PCM, which improved the performance of the BTMS. Further research found that a TEC connected with one side of the battery pack can improve the reliability of the system.

However, at a high discharge rate, the HP cannot sufficiently dissipate battery heat. Though the heat of the battery can be dissipated quickly by the cold side of TEC, the heat produced on the hot side is hard to be dissipated under natural conditions. Therefore, this paper proposes a BTMS combining an HP with TEC. With this system, the heat produced on the hot side of the TEC can be quickly transferred to the device dissipating heat by the HP. Both numerical simulations and discharging experiments were conducted to research the surface temperature rising characteristics of the battery with different discharge rates. To reduce the energy consumption but maintain the surface temperature of the battery within the optimal temperature, we performed experiments during which the TEC began to work at different temperatures.

2. Theoretical Analysis

A TAFEL-LAE895 100 Ah ternary Li-ion battery, which was produced by TAFEL company of Jiangsu province, China, was selected for analysis in this research. Its main parameters are provided in Table 1.

Table 1. TAFEL-LAE895 battery specifications.

Battery Model	Size (mm)	Capacity (V/Ah)	Maximum Discharge Rate (C)	Maximum Charge Rate (C)	Optimum Temperature (K)
TAFEL-LAE895	52 × 148 × 96 (L × W× H)	3.6/100	5	3	293–318

The heat generated by the battery includes electrochemical reaction heat, joule heat, polarized heat, and side reaction heat [23], which can be calculated by

$$Q = Q_r + Q_j + Q_p + Q_s \tag{1}$$

where Q is the total heat generated by the battery; Q_r is the electrochemical reaction heat, which refers to the heat generated by the electrochemical reaction when lithium ions are intercalated and deintercalated between the anode and cathode materials in the process of charging and discharging; Q_j represents joule heat generated by joule internal resistance; Q_p represents polarized heat, which refers to the heat generated by the polarized internal resistance due to the polarization of the battery during the charging and discharging process; and Q_s is side reaction heat, which is the heat generated in the process of the decomposition and reaction of the electrolyte, the thermal decomposition of the anode and cathode materials, and the decomposition of the separator.

Due to the influence of the environment and heat loss of transmission, directly measuring the heat generated by a Li-ion battery in the experiment was difficult. In this work, the calculation method proposed by Bernardi et al. [24] was adopted. Under the assumption that the heat generation inside the Li-ion battery is uniform, a theoretical formula of the heat generation rate was proposed [25,26]. The formula is simplified as

$$Q = I(E - U) - IT\frac{\partial E}{\partial T} \tag{2}$$

where I represents the charging and discharging current of the Li-ion battery; E and U are the open and closed circuit voltages, respectively; and T represents the surface temperature of the battery, the unit of which is kelvins. The first term on the right side of the equation represents the irreversible reaction heat, such as joule heat and electrochemically polarized heat, which can be represented by I^2R, where R is the total internal resistance of the Li-ion battery. The second term represents the heat generated by the reversible reaction, and the Li-ion battery generally takes a reference value of 0.042 [25]. After the values are inserted into the formula, we have

$$Q = I^2R - 0.042I$$

The value of R can be tested using a Hybrid Pulse Power Characteristic (HPPC) experiment. The value of R is calculated in Table 2.

Table 2. The value of R (total internal resistance of the Li-ion battery).

Discharge rate (C)	1	1.5	2	2.5	3
Value of R (Ω)	0.00157	0.00129	0.00115	0.00107	0.00104

The heat generated by the Li-ion battery at different discharge rates can be calculated as shown in Figure 1.

Note that the battery discharge rate is measured in C, which means that battery discharges its rated capacity within the specified time (C = current/rated capacity).

As shown in Figure 1, with increasing discharge rate, the heat-generating rate rises in an accelerated manner, from 11.5 W at a 1 C discharge rate to 80.7 W at a 3 C discharge rate.

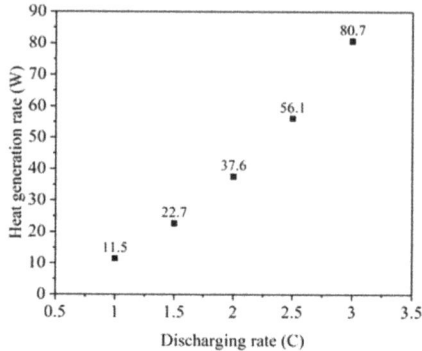

Figure 1. Theoretical rate of battery heat generation.

3. Numerical Simulations

Analyzing the temperature of a battery is a prerequisite for designing a highly efficient battery thermal management system. Numerical simulations were used to examine battery surface temperature change at different discharge rates to verify the effect of the theoretical model and help design the experimental BTMS model.

3.1. Simulation Model Design

3D design software, that is, Solidworks of Dassault Systemes S.A company, was used to build models of the battery, the BTMS based on an HP, and the combined HP and TEC system. The sizes of the models were designed as experimental devices. To achieve a more uniform temperature distribution and to accelerate heat dissipation, aluminum sheets were added to the surface of the battery, and grooved aluminum sheets were fixed to the HP. Heat sinks and fans were added. The model is shown in Figure 2.

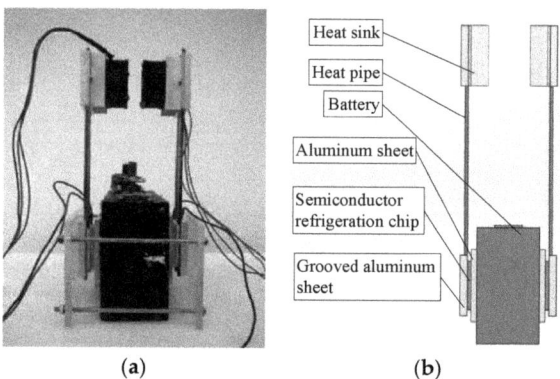

Figure 2. Battery thermal management system (BTMS) model: (**a**) experimental; (**b**) simulation models.

The 3D model was imported into Fluent, the computational fluid dynamics (CFD) software. For simulation, the following assumptions were made:

(1) Since the fluidity of the electrolyte inside the Li-ion battery is weak, we assumed that the heat transfer inside the battery could be ignored.
(2) The battery was considered as an ideal whole and its internal heat generation was uniform.

(3) The complex internal structure of the HP was regarded as a solid piece with equivalent thermal conductivity [27].

In the transient simulation model, first-order upwind and laminar flow method were adopted, and then the energy equation was turned on. The environmental temperature was set to ambient room temperature, which is 298 K. The boundary conditions of the battery were set to natural convection. The convective heat transfer coefficient was set to 3.5 $Wm^{-2}K^{-1}$, adding a wind speed of 7 $m\,s^{-1}$ to the heat sink to simulate the airflow speed generated by the fan. The heat sink was composed of aluminum, so the parameters of the heat sink, aluminum sheet, and grooved aluminum sheet were the same: density of 2719 $kg\,m^{-3}$, specific heat capacity of 871 $J\,kg^{-1}K^{-1}$, and thermal conductivity of 202.4 $Wm^{-1}K^{-1}$.

During the calculation, the residuals of the conservation equations, temperature of characteristics, and the characteristic plane were detected. When the residual was less than 10^{-6} and the temperature change at the detection point was less than 0.1%, the calculation was considered to reach convergence.

3.2. Simulation Results and Analysis

3.2.1. Simulation of Surface Temperature Rise of Battery in Natural State

For the natural working state of the battery pack, without any additional device to help dissipate heat, the model was meshed with 216,448 elements. The rate of generating heat is provided in Figure 1 and the simulation results of the battery pack are shown in Figure 3.

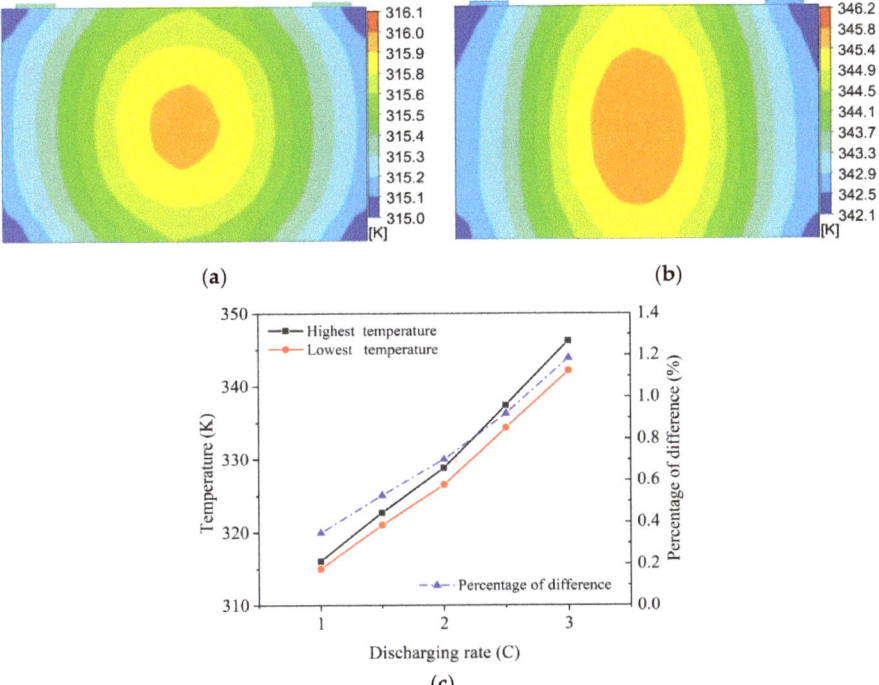

Figure 3. The surface temperature of the battery pack at different discharge rate in natural state: (**a**) temperature distribution at 1 C discharge rate; (**b**) temperature distribution at 3 C discharge rate and (**c**) surface temperature and ratio of difference.

Figure 3c shows the highest and lowest surface temperatures of the battery and the percentage of difference in surface temperature at the end of discharging. In Figure 3, with the increase in the discharge rate, the surface temperature shows an upward trend, as does the temperature difference ratio. The difference in temperature was moderate at less than 1.2%; thus, it was regarded as uniformly distributed. The temperature was 316.1 K at a 1 C discharge rate, which is within the range of optimal working temperature between 293 K to 318 K. However, at 2.5 and 3 C discharge rates, the surface temperatures all exceeded 323 K. Working for a long time at such a high temperature will have an irreversible impact on battery performance and cycle life. The heat generated by the battery will accumulate, causing thermal runaway and may lead to battery fire or explosion.

3.2.2. Simulation with an Added HP

With an HP, a heat sink, and a fan attached to the surface of the battery, the model was modeled with 292,605 elements with the same assumptions applied. The heat generation rate of the Li-ion battery selected in this paper was 80.7 W at the maximum 3 C discharge rate. Therefore, the HP with a power of 80 W, a length of 240 mm, a width of 11 mm, and a thickness of 2 mm was selected in this study. The wick material of the selected HP was copper, and the heat pipe was regarded as special copper with high thermal conductivity. The thermal conductivity of the HP is shown in Table 3, which was measured by an experimental method previously reported [27,28].

Table 3. The thermal conductivity of the heat pipe (HP).

Discharge rate (C)	1.5	2	2.5	3
Thermal conductivity (Wm^{-2}K^{-1})	5772	6937	7765	9006

The density and specific heat capacity of the HP were 4000 kg m^{-3} and 400 J kg^{-1}K^{-1}, respectively [27]. The results of the simulation are shown in Figure 4.

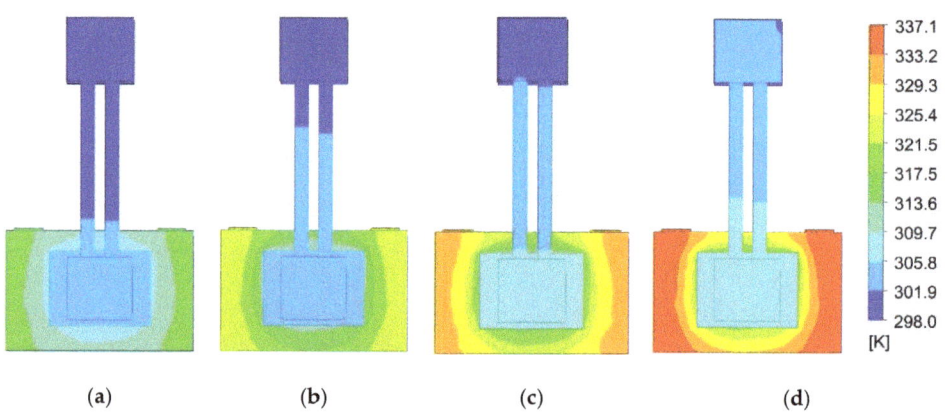

Figure 4. The surface temperature with different discharge rates based on the heat pipe HP model: (a) 1.5 C; (b) 2 C; (c) 2.5 C; (d) 3 C.

Figure 4 shows that the surface temperatures were significantly lower than in the previous natural state at different discharge rates. The maximum temperatures were 337.1 K at a 3 C discharge rate and 315 K at 1.5 C, which are 9.1 and 7.7 K lower than those in the natural state shown in Figure 3, respectively. However, the difference in surface temperature was substantially larger than that under the natural state because the width of the HP was narrower than that of the battery, and the surface heat of the battery near the HP was more quickly dissipated. The effectiveness of the HP was limited. At a

3 C discharge rates, the surface temperature still exceeded 323 K, indicating that additional measures were needed to further lower the temperature.

3.2.3. Simulations for Combined HP and TEC

As shown in Figure 5, the TEC was mainly composed of a series of N- and P-type thermo-elements, which were connected to each other using a metal conductor. When current passes through the unit, the energy transfers through the circuit and forms the cold and hot sides of the TEC [16].

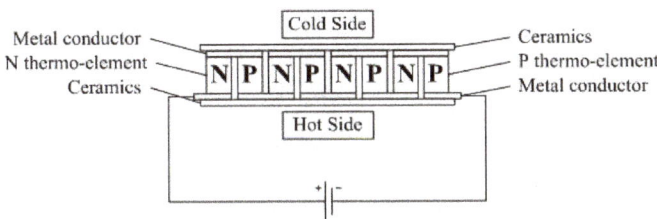

Figure 5. The thermoelectric cooling (TEC) structure.

When a TEC is used for refrigeration, the cold side is attached to the power source and the hot side is attached to the dissipating device. Therefore, TEC was regarded as a cuboid for simulation. The side of the cuboid with negative power simulated the cold side of the TEC to absorb heat, and the other side, with positive power, simulated the hot side to dissipate heat [29]. The power was calculated by the optimal working current, the corresponding voltage, and the coefficient of performance (COP). The optimal working current of the TEC selected was 3.96 A, and the corresponding voltage was 7.8 V. The power was 17.92 W = 3.96 A × 7.8 V × 0.58. The other part of the cuboid was regarded as particular material, whose thermal conductivity was 1.5 Wm^{-2}K^{-1}. With the added TEC, the model was meshed into 296,569 elements. The simulation results are shown in Figure 6.

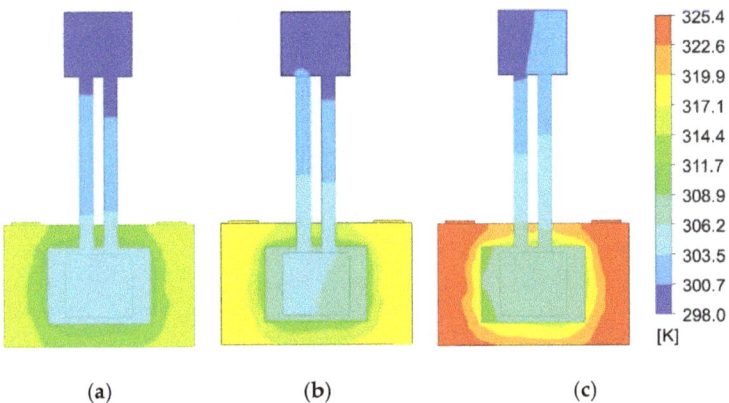

Figure 6. Surface temperatures at different discharge rates based on HP and TEC: (**a**) 2 C; (**b**) 2.5 C; (**c**) 3 C.

Figure 6 shows that the surface temperatures of the battery were significantly lower than that with only the HP at different discharge rates. The maximum temperature was 325.4 K at a 3 C discharge rate and 316.4 K at a 2 C discharge rate, which are 11.7 and 9 K lower, respectively, than that in the HP alone case shown in Figure 4. This combined BTMS can reduce the surface temperature of the battery pack and basically keep the battery working within a reasonable temperature range.

4. Experiment

The three discussed BTMS models analyzed by the simulations were tested to verify the practicality of the design and the simulation results.

4.1. Experimental Design

To avoid being influenced by the environmental temperature during the experiment, the battery pack was placed in a thermotank. The thermotank was set to 298 K to simulate the ambient environmental temperature; the battery pack was charged and discharged by a charging tester. The first constant current and then constant voltage (CC-CV) method was adopted to charge the battery. High accuracy and low-cost Omega T-type thermocouples were used as temperature sensors and were connected to a data acquisition instrument to acquire the temperature of the battery over time. A micro temperature control module was used to operate the TEC according to the surface temperature of the battery pack. Four Li-ion batteries were connected in series into a battery pack, and its working voltage was 11.2–16.8 V. Four temperature measuring points were selected on the front and back sides of each battery, and two more measuring points were arranged on opposite sides of each battery. Data obtained from the six measuring points were then averaged to represent the temperature characteristics of this battery. The measuring points of the battery are shown in Figure 7.

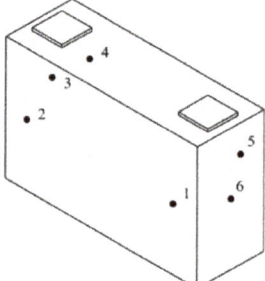

Figure 7. Placement of temperature sensors on a Li-ion cell.

4.2. BTMS Experiment and Results Analysis

4.2.1. Surface Temperature Rise of Battery in Natural State

Four batteries were connected in series with a distance of 2 Cm and placed in the thermotank for 1 h to stabilize the surface temperature of the battery at 298 K. Then, the experiments were conducted at different discharge rates. To ensure the same discharging capacity each time and avoid damage to the battery, a discharge capacity of 90 Ah was used. After each charging and discharging cycle, the battery was kept for 7 h in the thermotank to stabilize the surface temperature of the battery at 298 K again. The experimental results are shown in Figure 8.

As shown in Figure 8, the surface temperature of the battery continued to rise during the discharging period. The drop near the end of each curve indicates that the discharge was over and the battery had begun to cool down. The maximum surface temperature of the battery was 313.2 K at a 1 C discharge rate, and the battery worked within the optimal temperature range throughout the process. As the discharge rate increased, the discharging time shortened and the surface temperature of the battery increased. The maximum surface temperature of the battery reached 343 K at a 3 C discharge rate. A battery operating at this high temperature for a prolonged duration will undergo thermal runaway, which reduces battery capacity and shortens its service life. Therefore, this is not normally permitted and is the key reason to apply a BTMS to battery packs.

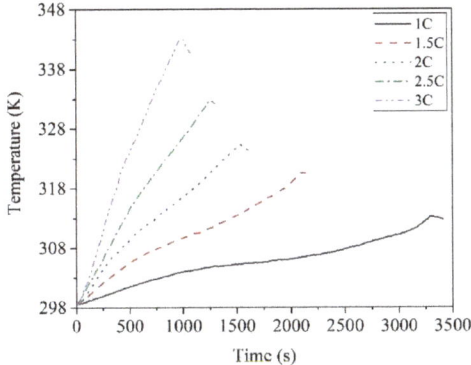

Figure 8. Surface temperature rise at different discharge rates in the natural state.

4.2.2. Surface Temperature Rise of Battery with BTMS and HP

Figure 9 shows the structure of the BTMS with the HP. One side of the aluminum sheet was in contact with the surface of the battery; the other side was in contact with the evaporation section of HP, which was fixed by a grooved aluminum sheet. The HP condensation section was connected with one side of the heat sink, and the other side of the heat sink with the fan.

Figure 9. The BTMS based on HP.

The environmental temperature was set to 298 K, and the fan started to work at 298 K. The experimental results are shown in Figure 10. With this system, we did not study the surface temperature rising characteristics of the battery at a 1 C discharge rate as the temperature remained within the optimal range without the HP.

As seen in Figure 10, the temperature for discharging at 1.5 C remained within the optimal range, and the maximum temperature was 319.7 K at a 2 C discharge rate and 328.2 K at a 2.5 C discharge rate, above the upper limit, but 5.6 and 4.6 K lower, respectively, than that in the natural state shown in Figure 8. Thus, it was necessary to take additional measures to further reduce temperature.

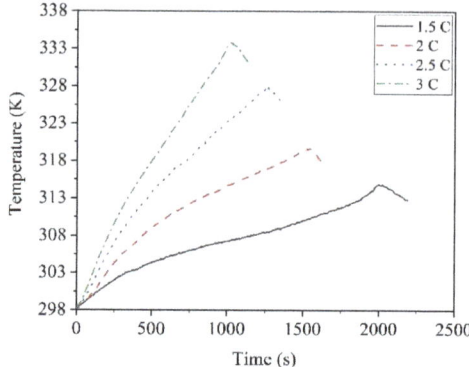

Figure 10. Surface temperature rise at different discharge rates in a BTMS based on an HP.

4.2.3. Surface Temperature Rise of Battery in BTMS Using Both HP and TEC

Figure 11 shows the structure with a TEC installed between the aluminum sheet and the HP. The cold side of the TEC was connected to the aluminum sheet, which can accelerate heat loss. The hot side was connected to the HP, and the heat generated by the hot side of TEC was transmitted to the heat sink through the HP, and then the fan accelerated the loss of heat.

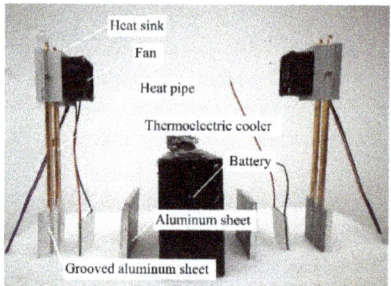

Figure 11. The BTMS based on an HP and TEC.

The two kinds of limits on the working conditions of the TEC are the maximum cooling capacity and the maximum COP. The maximum cooling capacity is when the TEC works at maximum current, but then it consumes a larger amount of power and the obtained COP is also smaller. Although the working condition at maximum COP is better economically as it consumes less power, the cooling capacity is very low. Therefore, the optimal working current must be selected to increase the cooling capacity of the TEC and consume less power. The TEC used in this study was TCE1-12705, which has an optimal working current of 3.96 A according to Wang [19]. The environmental temperature was set to 298 K, and the TEC and the fan begin to work at 298 K. The experimental results are shown in Figure 12.

Figure 12 shows that at 2 C and 2.5 C discharge rates, the maximum surface temperature of the battery reached 313.5 K and 316.6 K, respectively, within the optimal working temperature range. At 3 C, the surface temperature of the battery reached 318 K in about 800 s, and 322 K at the end of discharging, which is 11.8 K lower than that of the BTMS based on an HP alone, shown in Figure 10. According to the experimental results analyzed above, this combined BTMS can effectively control the surface temperature of the battery pack.

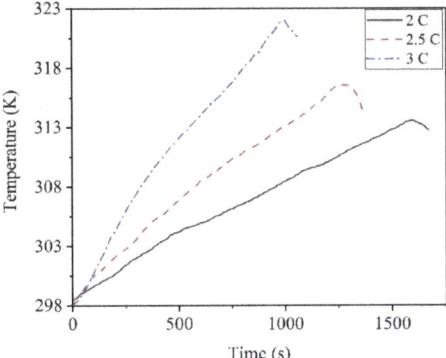

Figure 12. Surface temperature rise at different discharge rates in BTMS based on HP and TEC.

4.3. Optimal Working Temperature of TEC

When the BTMS based on HP and TEC was applied to the battery, the surface temperature of the battery was lower than 318 K at the 2 C and 2.5 C discharge rates. To reduce energy consumption, the optimal working temperature of the TEC was selected, and the surface temperature of the battery was still under 318 K. The environmental temperature was 298 K, and the upper limit of the optimal working temperature of the battery was 318 K. Therefore, the TEC worked when the surface temperature of battery reached 298 K, 303 K, 308 K, 313 K, and 318 K. At the 3 C discharge rate, the surface temperature of the battery exceeded 318 K, and we did not study the surface temperature rise characteristics of the battery at the 3 C discharge rate.

As the environmental temperature was set to 298 K, the fan was turned on at 298 K, and the TEC began to work at 298 K, 303 K, 308 K, 313 K, and 318 K. The experimental results of the discharge rates of 2 C and 2.5 C are shown in Figure 13.

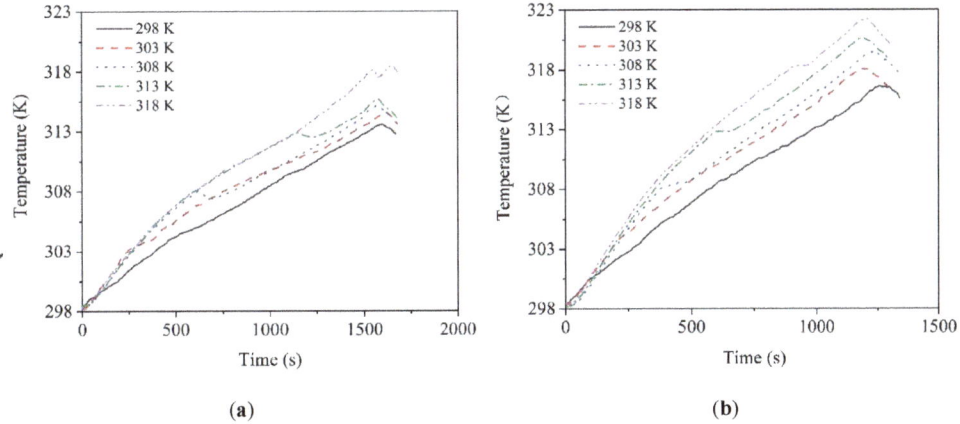

Figure 13. Surface temperature rise with the TEC beginning to work at different temperatures: (a) 2 C; (b) 2.5 C.

As shown in Figure 13, with increasing TEC working temperature, the surface temperature of the battery also rose. When the TEC began to work at 308 K, 313 K, and 318 K, the temperature curve tended to decrease at the start of TEC working. At that moment, the cold side of the TEC accelerated the loss of surface temperature of the battery, and the heat generated from the hot side was transferred to the heat sink through the HP.

At the 2 C discharge rate, when the TEC began to work at 318 K, the surface temperature of the battery remained near 318 K until discharge was over. At a 2.5 C discharge rate, when the TEC began to work at 303 K, the surface temperature of the battery remained near 318 K until discharge was over. Therefore, the optimal working temperature for TEC to begin working is 318 K at a 2 C discharge rate and 303 K at a 2.5 C discharge rate.

5. Discussion and Conclusions

5.1. Discussion

5.1.1. Comparison of Simulation and Experimental Results

Based on the different BTM models, simulations and experiments were conducted to research temperature increase characteristics of a battery at different discharge rates. The maximum battery surface temperatures at different discharge rates are shown in Figure 14.

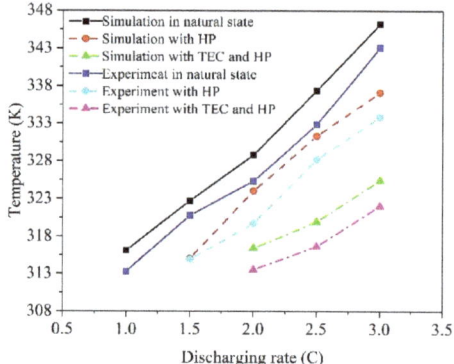

Figure 14. Comparison of maximum temperatures between simulation and experimental results.

As seen in Figure 14, with increasing discharge rate, the battery surface temperature increased. For the same model under the same discharge rate, the surface temperature was the highest in the natural state and the lowest with the combined HP and TEC, and the maximum surface temperature measured in the experiment was lower than in the simulation, which is reasonable. The maximum difference in surface temperature between the experiment and simulation was within 5 K, showing a very good agreement between the two approaches, verifying the simulated results.

5.1.2. Comparison of Battery Surface Temperature Increase Rate

The rising rates of battery temperature directly reflect the change in temperature. The temperature rising rates of the battery are shown in Figure 15.

Under the same discharge rate, the rising rate of the battery surface temperature was the highest in the natural state and the lowest with the BTMS based on combined HP and TEC. The progressive improvement in reducing battery surface temperature demonstrated the effectiveness of the designed model.

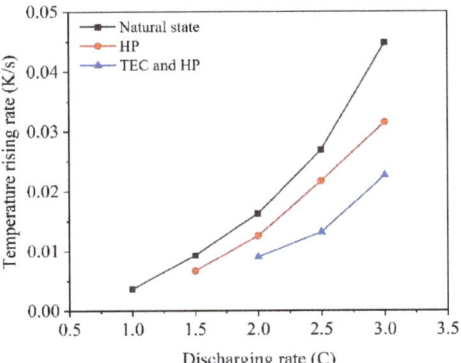

Figure 15. Battery surface temperature rising rates.

5.2. Conclusions

As the discharge rate increases, the surface temperature of a battery rises. We designed a battery thermal management system that combined HP and TEC and investigated its performance. Simulations and experiments were conducted to study the surface temperature rising characteristics of the battery with the system at different discharge rates. Based on TAFEL-LAE895 100 Ah ternary lithium ion batteries, the conclusions were drawn as follows.

1. Compared with the natural state, a BTMS is needed to maintain the operational temperature of batteries within the optimal range to maintain performance output, reduce damage, and prolong the service life.
2. The battery thermal management system studied, which is based on a system with HP and TEC, can effectively reduce the surface temperature rising rate at different discharge rates and maintain the surface temperature broadly within the optimal range of the battery. However, the difference in surface temperature is larger with a BTMS than in the natural state, which can be further studied in the future.
3. For the battery selected in this study, some specific design guidance can be provided for the optimal performance of the battery: there is no need to introduce heat dissipation measures for a discharge rate of 1 C; a BTMS based on an HP is needed at a 1.5 C discharge rate; a BTMS based on HP and TEC is required for discharging at 2 C, 2.5 C, and 3 C; and the discharging duration at a 3 C discharge rate must be controlled.
4. To reduce energy consumption and keep surface temperature of the battery within 318 K, when the BTMS based on TEC and HP was applied to the battery selected in this study, the optimal working temperature for the TEC to begin to work was 318 K at 2 C a discharge rate and 303 K at a 2.5 C discharge rate.

Author Contributions: Conceptualization, C.Z.; methodology, H.G.; software, S.Z. and K.L.; formal analysis, Z.X. and S.C.; data curation, B.W. and Z.X.; writing—original draft preparation, Z.X.; writing—review and editing, H.G. and Z.X. All authors have read and agreed to the published version of the manuscript.

Funding: This research was funded by the Natural Science Foundation of China, grant number 51974229" and "The APC was funded by Xi'an University of Science and Technology.

Acknowledgments: This study was supported by a research grant from the Natural Science Foundation of China (Grant No. 51974229).

Conflicts of Interest: Declare no conflicts of interest.Authors must identify and declare any personal circumstances or interest that may be perceived as inappropriately influencing the representation or interpretation of reported research results. The funders had no role in the design of the study; in the collection, analyses, or interpretation of data; in the writing of the manuscript, or in the decision to publish the results.

References

1. Rao, Z.; Wang, Q.; Huang, C. Investigation of the thermal performance of phase change material/mini-channel coupled battery thermal management system. *J. Appl. Energy* **2016**, *164*, 659–669. [CrossRef]
2. Hannan, M.A.; Lipu, M.H.; Hussain, A.; Mohamed, A. A review of lithium-ion battery state of charge estimation and management system in electric vehicle applications: Challenges and recommendations. *J. Renew. Sustain. Energy Rev.* **2017**, *78*, 834–854. [CrossRef]
3. Widmaier, M.; Jäckel, N.; Zeiger, M.; Abuzarli, M.; Engel, C.; Bommer, L.; Presser, V. Influence of carbon distribution on the electrochemical performance and stability of lithium titanate based energy storage devices. *J. Electrochim. Acta* **2017**, *247*, 1006–1018. [CrossRef]
4. He, F.; Ma, L. Thermal management of batteries employing active temperature control and reciprocating cooling flow. *J. Int. J. Heat Mass Transfer* **2015**, *83*, 164–172. [CrossRef]
5. Greco, A.; Cao, D.; Jiang, X.; Yang, H. A theoretical and computational study of lithium-ion battery thermal management for electric vehicles using heat pipe. *J. Power Sources* **2014**, *257*, 344–355. [CrossRef]
6. Feng, X.; Lu, L.; Ouyang, M.; Li, J.; He, X. A 3D thermal runaway propagation model for a large format lithium ion battery module. *J. Energy* **2016**, *115*, 194–208. [CrossRef]
7. Zhang, G.; Zhang, H. Progress in application of phase change materials in battery module thermal management system. *J. Mater. Rev.* **2006**, *20*, 9–12.
8. Liang, J.; Gan, Y.; Li, Y. Investigation on the thermal performance of a battery thermal management system using heat pipe under different ambient temperatures. *J. Energy Convers. Manag.* **2018**, *155*, 1–9. [CrossRef]
9. Rao, Z.; Qian, Z.; Kuang, Y.; Li, Y. Thermal performance of liquid cooling based thermal management system for cylindrical lithium-ion battery module with variable contact surface. *J. Appl. Therm. Eng.* **2017**, *123*, 1514–1522. [CrossRef]
10. Pesaran, A.; Vlahinos, A.; Stuart, T. Cooling and preheating of batteries in hybrid electric vehicles. In Proceedings of the 6th ASME-JSME Thermal Engineering Joint Conference, Hawaii Island, HI, USA, 16–20 March 2003; pp. 1–7.
11. Sabbah, R.; Kizilel, R.; Selman, J.R.; Al-Hallaj, S. Active (air-cooled) vs. passive (phase change material) thermal management of high power lithium-ion packs: Limitation of temperature rise and uniformity of temperature distribution. *J. Power Sources* **2008**, *182*, 630–638. [CrossRef]
12. Zhang, G.; Rao, Z.; Wu, Z.; Fu, L. Experimental investigation on the heat dissipation effect of power battery pack cooled with phase change materials. *J. Chem. Ind. Eng. Prog.* **2009**, *28*, 23–26+40.
13. Rao, Z.; Zhang, G. Thermal properties of paraffin wax-based composites containing graphite. *J. Energy Sources Part A Recovery Util. Environ. Eff.* **2011**, *33*, 587–593. [CrossRef]
14. Rao, Z.; Wu, Z.; Zhang, G. Power Cell Device with Phase-Change Material Cooling System. Chinese Patent ZL200920055746.7, 5 May 2010.
15. Wu, W.; Yang, X.; Zhang, G. An experimental study of thermal management system using copper mesh-enhanced composite phase change materials for power battery pack. *J. Energy* **2016**, *113*, 909–916. [CrossRef]
16. Arora, S. Selection of thermal management system for modular battery packs of electric vehicle: A review of existing and emerging technologies. *J. Power Source* **2018**, *400*, 621–640. [CrossRef]
17. Brinkmann, R.; Radt, B.; Flamm, C.; Kampmeier, J.; Koop, N.; Birngruber, R. Influence of temperature and time on thermally induced forces in corneal collagen and the effect on laser thermokeratoplasty. *J. Cataract Refract. Surg.* **2000**, *26*, 744–754. [CrossRef]
18. Wang, Y. *Study on the Performance of TEC and Its Application in Power Battery Thermal Management*; Guangdong University of Technology: Guangzhou, China, 2015.
19. Smith, J.; Singh, R.; Hinterberger, M.; Mochizuki, M. Battery thermal management system for electric vehicle using heat pipes. *J. Int. J. Therm. Sci.* **2018**, *134*, 517–529. [CrossRef]
20. Deng, S.; Li, K.; Xie, Y.; Wu, C.; Wang, P.; Yu, M.; Zheng, J. Heat Pipe Thermal Managemet Based on High-Rate Discharge and Pulse Cycle Tests for Lithium-Ion Batteries. *J. Energy* **2019**, *12*, 3143.
21. Lu, G.; Han, H.; Yang, D.; Ren, K. Research on thermal management application of power battery box based on semiconductor refrigeration technology. *J. Electron. World* **2014**, *3*, 186–188.
22. Arora, S.; Kapoor, A.; Shen, W. A novel thermal management system for improving discharge/charge performance of Li-ion battery packs under abuse. *J. Power Source* **2018**, *378*, 759–775. [CrossRef]

23. Wang, T. *Study on Layered Air Cooling Thermal Management for Lithium-ion Battery Pack*; Beijing University of Technology: Beijing, China, 2016.
24. Bernardi, D.; Pawlikowski, E.; Newman, J. A general energy balance for battery systems. *J. Electrochem. Soc.* **1985**, *132*, 5–12. [CrossRef]
25. Zhang, C.W.; Chen, S.R.; Gao, H.B.; Xu, K.J.; Xia, Z.; Li, S.T. Study of Thermal Management System Using Composite Phase Change Materials and Thermoelelectric Cooling Sheet for Power Battery Pack. *J. Energy* **2019**, *12*, 1937.
26. Worwood, D.; Marco, J.; Kellner, Q.; Hosseinzadeh, E.; McGlen, R.; Mullen, D.; Greenwood, D. Experimental Analysis of a Novel Cooling Material for Large Format Automotive Lithium-Ion Cells. *J. Energy* **2019**, *12*, 1251. [CrossRef]
27. Wang, J.; Gan, Y.; Liang, J.; Tan, M.; Li, Y. Sensitivity analysis of factors influencing a heat pipe-based thermal management system for a battery module with cylindrical cells. *J. Appl. Therm. Eng.* **2019**, *151*, 475–485. [CrossRef]
28. Li, Y.; He, H.; Zeng, Z. Evaporation and condensation heat transfer in a heat pipe with a sintered-grooved composite wick. *J. Appl. Therm. Eng.* **2013**, *50*, 342–351. [CrossRef]
29. Wang, Y.; Yin, B.; Qi, C. Analysis of Study Heat Dissipation Application to Electronic Components Based on Thermoelectric Cooling. *J. Electro Mech. Eng.* **2017**, *33*, 47–51.

© 2020 by the authors. Licensee MDPI, Basel, Switzerland. This article is an open access article distributed under the terms and conditions of the Creative Commons Attribution (CC BY) license (http://creativecommons.org/licenses/by/4.0/).

Article

Heat Transport Capacity of an Axial-Rotating Single-Loop Oscillating Heat Pipe for Abrasive-Milling Tools

Ning Qian [1], Yucan Fu [1,*], Marco Marengo [2], Jiuhua Xu [1], Jiajia Chen [3] and Fan Jiang [1]

[1] College of Mechanical and Electrical Engineering, Nanjing University of Aeronautics and Astronautics, Nanjing 210016, China
[2] Advanced Engineering Centre, University of Brighton, Brighton BN2 4GJ, UK
[3] College of Mechanical and Electronic Engineering, Nanjing Forestry University, Nanjing 210037, China
* Correspondence: yucanfu@nuaa.edu.cn; Tel.: +86-(0)25-84895857

Received: 25 March 2020; Accepted: 7 April 2020; Published: 30 April 2020

Abstract: In order to enhance heat transfer in the abrasive-milling processes to reduce thermal damage, the concept of employing oscillating heat pipes (OHPs) in an abrasive-milling tool is proposed. A single-loop OHP (SLOHP) is positioned on the plane parallel to the rotational axis of the tool. In this case, centrifugal accelerations do not segregate the fluid between the evaporator and condenser. The experimental investigation is conducted to study the effects of centrifugal acceleration (0–738 m/s^2), heat flux (9100–31,850 W/m^2) and working fluids (methanol, acetone and water) on the thermal performance. Results show that the centrifugal acceleration has a positive influence on the thermal performance of the axial-rotating SLOHP when filled with acetone or methanol. As for water, with the increase of centrifugal acceleration, the heat transfer performance first increases and then decreases. The thermal performance enhances for higher heat flux rises for all the fluids. The flow inside the axial-rotating SLOHP is analyzed by a slow-motion visualization supported by the theoretical analysis. Based on the theoretical analysis, the rotation will increase the resistance for the vapor to penetrate through the liquid slugs to form an annular flow, which is verified by the visualization.

Keywords: oscillating heat pipes; heat transfer; milling cooling; abrasive-milling processes

1. Introduction

During abrasive-milling processes, a large amount of heat is generated in the contact zone between the abrasive-milling tool and workpiece. Consequently, when abrasive-milling hard-to-machine materials with low heat conduction (e.g., nickel-based superalloy, carbon fiber reinforced composites or ceramic matrix composites), the heat gathers in contact zone and leads to critical hot spots and serious thermal damage for both the workpiece and tool [1–4]. To solve this problem, the heat should be transferred out through the tool in time to reduce the stored heat, lowering the temperature in the contact zone, with the result decreasing the risk of thermal damage.

Heat pipes are passive heat transfer devices with excellent heat transport capacity. They are widely used in the microelectronics, manufacturing, and automobile and aerospace industries for thermal management. Lately, heat pipes have been applied to the machining process to enhance heat transfer [5–9]. The heat is absorbed by the working fluid in the evaporator of the heat pipe, which, in the case of the tool, is near the contact zone. The phase change occurs, and vapor moves to the condenser to release heat by condensing into liquid. After this, the liquid returns to the evaporator either by capillary wick or unbalanced pressure distribution. In the case of applying heat pipes in the cutting tools, such fluid motion continues to transfer out the heat, with high heat transport capacity during the machining processes. The combination of heat pipes and cutting tools was conducted by several researchers [5–7], and the heat pipe cutting tool shows a great advantage in enhancing the heat and cooling the tool in the processes. He et al. [8,9] and Chen et al. [10,11] designed and produced axial-rotating and radial-rotating heat pipe grinding wheels. A titanium alloy and nickel-based superalloy were successfully dry-grinded by heat pipe grinding wheels without grinding burnout. An in-depth numerical analysis was conducted, where results show that the wickless heat pipe can assist heat transfer under a low rotational speed. Nucleate boiling was found to be totally suppressed while increasing the centrifugal acceleration. Laminar convection heat transfer is the heat transfer mode in the evaporator when the centrifugal acceleration is more than 1000 times the gravity acceleration.

The oscillating heat pipe (OHP, also called the pulsating heat pipe) is a novel heat transfer device, which has been developed since the 1990s. It is made of a meandering capillary tube which is partially filled with working fluid [12]. In recent years, experimental and theoretical research has been carried out on the thermal performance of OHPs and their heat transfer mechanism. The heat transport capacity of OHPs is affected by geometric, physical and operational parameters, such as inner diameters, working fluid properties, filling ratio, etc. It was found that within the critical inner diameter, an OHP with a larger inner diameter has a better heat transfer capability [13]. The influence of the working fluid on the heat transport capacity of the OHP has been described by several researchers [14,15]. The thermal performance of an OHP is also determined by its flow regimes, which in turn are affected by the working fluids. In general, fluids with lower dynamic viscosity and smaller surface tension lead to better performance, due to their lower friction and the higher wettability, which lead to a higher response to the pressure variation and the presence of a liquid film on the wall (decreasing the probability of dry spot occurrence). Fluids with a lower latent heat (such as acetone) bring powerful oscillations into the OHP, while they have a higher possibility of dry-out. OHPs charged with fluids with higher latent heat (e.g., water) may have better heat transfer performance in terms of maximum heat flux, but the start-up power is higher. The influence of operational parameters on the heat transfer was investigated by some researchers, such as Stevens et al. [16]. Operational parameters, which include a filling ratio (FR), inclination angle and heat load, etc., have significant effects on the heat transfer. In most of the experiments, there exists an optimized filling ratio which is within 40% to 60%, and the OHP will have a better heat transfer ability under the inclination of 50° to 90° for the proposed four turns closed-loop OHP [13,14,17,18]. Below the maximum heat flux, increasing the heat load, the flow regime develops from a bubbly flow to a slug flow to a transient flow, and then to an annular flow. As a result, the heat transfer performance varies depending on the flow patterns [19,20].

Due to the high heat transfer capability and the simple wickless structure, OHPs have the potential for application in the machining process for the enhancement of heat transfer. The OHP cutting tool was first proposed by Wu et al. [21,22]. A Ti-6Al-4V alloy was dry-cut by an OHP cutting tool. Compared with conventional cutting tools, the cutting tool with the OHP reduced the process temperature, and the tool wear-rate became slower by 20%. As for the grinding process, Qian et al. [19] introduced the concept of the OHP grinding wheel. The heat transfer analysis was studied to the point of the proof of concept. The heat transfer prediction model was built, and the start-up performance was also analyzed [23,24].

In order to dissipate the heat generated in the contact zone during the abrasive-milling of hard-to-machine materials, the concept of combining the OHP with the abrasive-milling tool is proposed, as illustrated in Figure 1. Several single-loop OHPs are formed by the independent channels machined in the tool matrix. The generated heat in the contact zone is transferred into the evaporator through the abrasive layers, and is then brought to the condenser to transfer out.

In this case, a rotation is involved. Experimental investigations of the heat transfer and flow patterns of "flower-shaped" radial-rotating OHPs were conducted by Aboutalebi et al. [25], Ebrahimi et al. [26], Liou et al. [27] and On-ai et al. [28]. In their studies, the maximum rotational speed reached 800 revolutions per minute (rpm), and the maximum centrifugal acceleration reached around 20 times gravity acceleration. For such a range of rotational speed and centrifugal acceleration, they all found that employing rotational speed was a way to enhance the internal flow, and the centrifugal acceleration has a positive influence on the increase of thermal performance. These oscillating heat pipes rotate around the central axis of the tool with the condenser at the centre and the evaporator at the external side. In this case, centrifugal accelerations segregate the fluid between the evaporator and condenser, pushing the fluid to the evaporator.

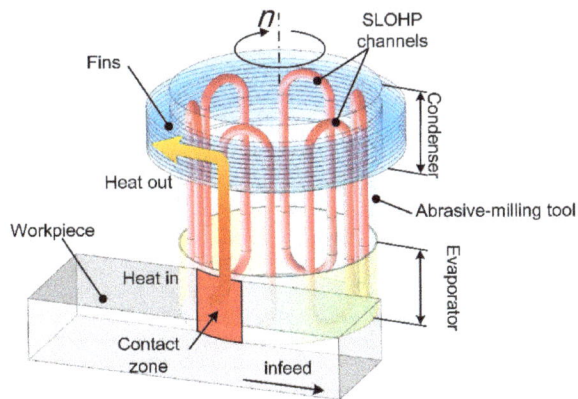

Figure 1. Illustration of an OHP abrasive-milling tool.

In our case, the OHP, whose central axis is aligned to the rotational axis of the tool, is positioned on the plane parallel to the rotational axis of the tool. In this case, centrifugal accelerations do not segregate the fluid between the evaporator and condenser. To the best of our knowledge, there is no previous paper on the influence of rotation on such positioning of OHPs, especially for a single-loop OHP.

In this paper, experiments are conducted to investigate the effects of centrifugal acceleration, heat flux and working fluids on the thermal performance of OHPs under axial-rotating conditions. The flow of liquid slugs and vapor plugs inside the axial-rotating SLOHP is also analyzed by the theoretical method. Moreover, the flow is also observed through a slow-motion camera.

2. Experimental Preparation and Data Processing

2.1. Description of Experimental Apparatus

The oscillating heat pipe (OHP) abrasive-milling tool (OHP tool) comprises several axial-rotating single-loop oscillating heat pipes (SLOHPs). For simplification, in this paper, one axial-rotating SLOHP is designed and made for the experiment, as shown in Figure 2. The axial-rotating OHP includes a holder, a copper SLOHP, a slip ring, thermocouples and Ni-Cr heating wires. The SLOHP is mounted on the holder 30 mm away from its rotational axis, which is the same as the radius of the OHP tool. The outer and inner diameters of a SLOHP are 5 mm and 3 mm, respectively. The evaporator, adiabatic section and condenser are 30 mm, 40 mm and 30 mm long, respectively. The SLOHP is evacuated to the pressure of 10^{-2} Pa, and then filled with acetone (CH_3COCH_3), methanol (CH_3OH) or deionized water (DI water) as working fluids, with a filling ratio of 55%, through the vacuum and injection tube. The axial-rotating SLOHP is bottom-heated by a Ni-Cr wire, which is connected to the KIKUSUI PZB40-10 power supply (limits of error: ±0.1 W) through the slip ring and copper brush. The axial-rotating SLOHP is wrapped with an aerogel insulation to avoid heat leak. The condenser is cooled by the 0.4 MPa cold air jet at a temperature under 5 °C. Temperatures of the evaporator and condenser are measured by Omega type-K thermocouples (limits of error: 0.4%). Signals are transferred through the slip ring and copper brush to the NI-USB 6366 card under the sampling rate of 10 Hz, and signals are processed by NI LabView and NI DIAdem software (National Instruments,

Shanghai, China). The environment temperature is maintained at 25±1° C. The setup of the experiment is illustrated in Figure 2.

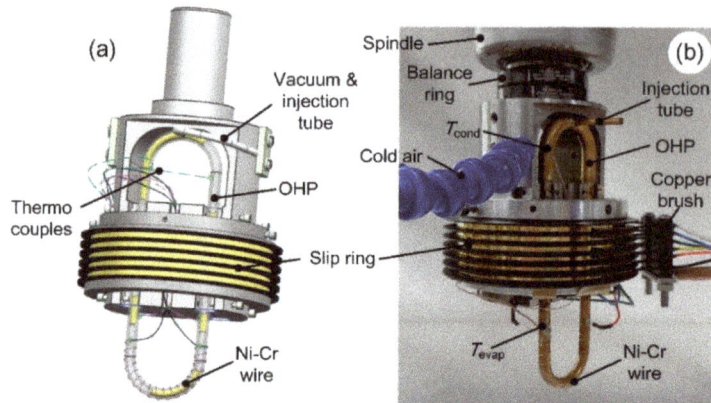

Figure 2. Experimental preparation of axial-rotating oscillating heat pipe: (**a**) illustration of axial-rotating OHP; and (**b**) experimental setup (aerogel insulation was removed for better illustration).

In the abrasive-milling process, the rotational speed of the tool varies from several hundred to a thousand rpm, and the effective heat flux generated in the contact zone is around 10^5 W/m^2. In this case, in order to study the thermal performance of the axial-rotating SLOHP under the same conditions with the abrasive-milling process, the rotational speeds in the experiment are 0, 300, 950, and 1500 rpm, with a relative centrifugal acceleration of 0, 30, 296, and 738 m/s^2. The heat flux used in the experiment varies from 9100 W/m^2 to 31,850 W/m^2, with an increment of 4550 W/m^2. The detailed conditions are shown in Table 1.

Table 1. Experimental conditions.

Conditions	Values
Rotational speed (rpm)	0, 300, 950, 1500
Centrifugal acceleration (m/s^2)	0 (0 g), 30 (3.1 g), 296 (30.2 g), 738 (75.3 g)
Heat flux (W/m^2)	9100, 13,650, 18,200, 22,750, 27,300, 31,850
Working fluids	Acetone, DI water, methanol
Filling ratio	55%
Heating mode	Bottom-heated in vertical

The thermophysical properties of working fluids (i.e., methanol, acetone and DI water) are shown in Table 2.

Table 2. Physical properties of working fluids at 1 atm.

Fluid	Boiling Point (°C)	Density (kg/m^3)	Specific Heat (kJ/kg °C)	Thermal Conductivity (W/m·K)	Latent Heat (kJ/kg)	$(dp/dT)_{sat}$ (kPa/°C)	Dynamic Viscosity (mPa·s)	Surface Tension (mN/m)
DI water	100.0	998	4.18	0.599	2257	1.30	1.01	72.8
Acetone	56.2	792	2.35	0.170	523	3.10	0.32	23.7
Methanol	64.7	791	2.48	0.212	1101	3.55	0.6	22.6

2.2. Data Processing

Thermal resistance is used as the measurement of the thermal performance of the axial-rotating SLOHP. The thermal resistant is defined as follows

$$R = \frac{Q}{T_{evap} - T_{cond}},\tag{1}$$

where Q is the heat load, and T_{evap} and T_{cond} are the time-averaged temperatures in evaporator and condenser. Lower thermal resistance means a better thermal performance of the axial-rotating SLOHP. The uncertainty of the thermal resistance is evaluated based on the method in [19]. The uncertainty is less than 5%.

3. Flow Visualization and Modeling

3.1. Flow Inside Axial-Rotating Single-Loop Oscillating Heat Pipe (SLOHP) Under Rotation

A glass axial-rotating SLOHP is made for visualization. The flow of working fluid inside the axial-rotating SLOHP is captured by a portable camera at a 240-frames-per-second speed (eight times slower). Due to the limitation of the apparatus and in the light of safety, the rotational speed is set up to 300 rpm in this case.

Under the static state and heat flux of 22,750 W/m^2, the flow pattern of the working fluid, acetone, is a veering circulation with an annular flow and slug flow. When the vapor bubbles and plugs generate, expand and move up to the condenser, the vapor easily penetrates the liquid slugs to form the annular flow (see Figure 3a(i–iii)), while the flow in the neighboring tube remains a train of liquid slugs and vapor plugs (see Figure 3a(ii,iii)). Besides, the direction of circulation changes from time to time, as shown in Figure 3a(ii,iii).

When the axial-rotation is applied, as mentioned in the theoretical analysis in Section 3.2, the resistance for the vapor to penetrate the liquid slug becomes larger, and as a result it is more difficult for the vapor to penetrate through the liquid slug to form the annular flow. The flows inside the axial-rotating SLOHP at the speeds of 60, 150 and 300 rpm are shown in Figure 3b–d. At the rotating speed of 60 rpm, the flow is a veering slug flow, which generates a train of liquid slugs and vapor plugs (see Figure 3b(ii)). Sometimes, when a train of liquid slugs and vapor plugs moves from evaporator to condenser, it will slow down and stop, and a new train of liquid slugs and vapor plugs is generated in the neighboring tube and moves upwards, causing the flow direction to change (see Figure 3b(ii,iii)). The whole process is illustrated in Figure 3b. At the rotating speeds of 150 and 300 rpm, the unidirectional circulation forms. Since the SLOHP rotates clockwise, the inertia causes the working fluid to flow in a one-way, clockwise direction inside the SLOHP. In this case, liquid slugs generate and move up to the condenser in the left tube, while in the right tube, liquid slugs and vapor plugs oscillate at the adiabatic section (see Figure 3c(i,iii) and Figure 3d(ii,iv)). Though flows at rotational speeds of 150 rpm and 300 rpm are similar, new trains of liquid slugs and vapor plugs are generated at the speed of 150 rpm (Figure 3c(i,iii)), while just one liquid slug forms and moves up at a time at the speed of 300 rpm (Figure 3d(ii,iv)).

As mentioned before, under the static state, vapor easily penetrates through the liquid slugs to form an annular flow. Under axial-rotation, especially when the speed exceeds 150 rpm, unidirectional circulation forms. It takes a relatively long time to form a long train of liquid slugs and vapor plugs at the speed of 150 rpm. Nevertheless, one short liquid slug is generated at a shorter time at the speed of 300 rpm, as shown in Figure 4.

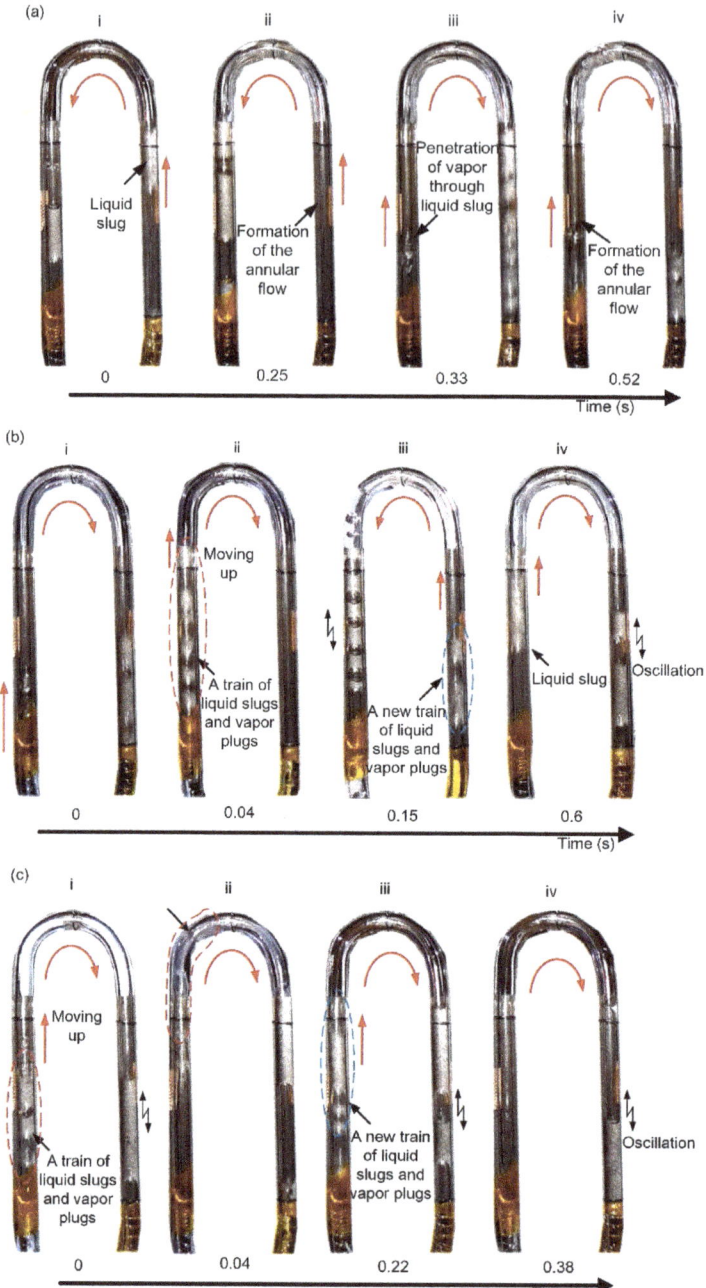

Figure 3. Cont.

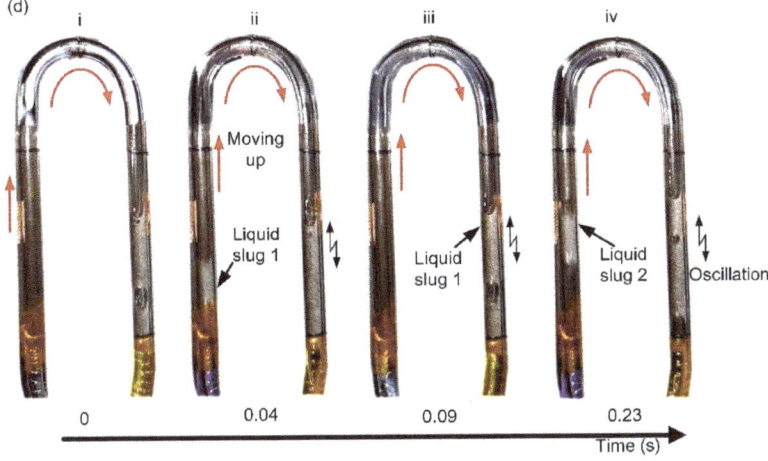

Figure 3. Flow inside the axial-rotating SLOHP filled with acetone under 22,750 W/m²: (**a**) at static state; (**b**) at 60 rpm rotating speed; (**c**) at 150 rpm rotating speed; and (**d**) at 300 rpm rotating speed.

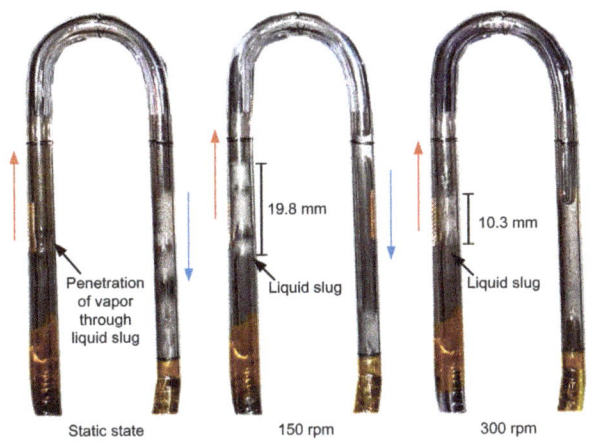

Figure 4. Comparison of liquid slug lengths under different rotating speeds (heat flux of 22,750 W/m²).

3.2. Modelling of Flow Under Rotation

For a well-working oscillating heat pipe (OHP), a train of liquid slugs and vapor plugs must form and exist in the system. At the moment that the evaporator is heated, the liquid turns into vapor, causing the expansion of the vapor volume. When the vapor velocity exceeds some critical value, the vapor plug penetrates all liquid plugs and generates an annular flow. At this moment, the mass-spring system, consisting of a train of liquid slugs and vapor plugs, vanishes. The oscillating motion stops, and the OHP reaches the highest heat transport capacity [29]. The illustration of the flow inside the OHP is shown in Figure 5.

When the vapor penetrates the liquid slug, the momentum that is generated by the vapor plugs will be used to overcome the total forces applying on the liquid plug, i.e.,

$$\int_0^{r_0} 2\pi \rho_v u_v^2 r dr = (-p_1 + p_2)\pi r_0^2 + 2\pi r_0 \sigma (cos\alpha_r + cos\alpha_a) + 2\pi r_0 \left(\int_0^{L_l} \tau_w dx \right), \quad (2)$$

where ρ_v is the vapor density, p_1 and p_2 are the vapor pressure, r_0 is the radius of the tube, σ is the surface tension, α_a and α_r are the advancing and receding contact angles and τ_w is the shear stress. The left term is the momentum produced by the vapor due to the vapor volume expansion. The right of the equation is marked as the total resistant force. The first term of the right part is the pressure differences acting on the liquid slug and vapor plug, while the second term of the right part is due to the surface tensions and the last term is due to the shear stress between the wall and fluid.

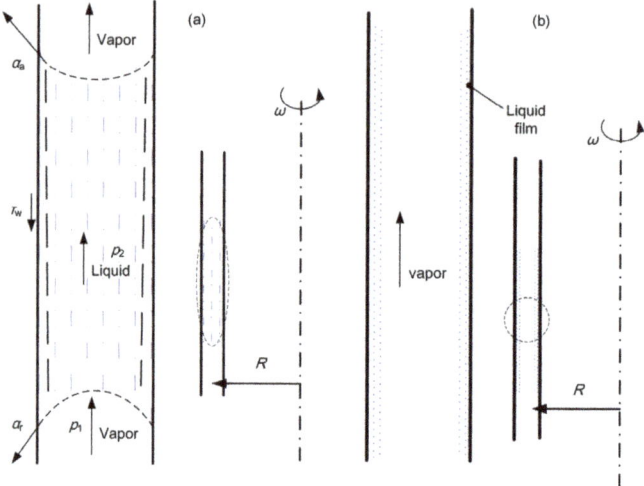

Figure 5. Illustration of liquid slug moving in the OHP: (**a**) right before the vapor penetrating through the liquid slug; and (**b**) right after the penetration of vapor through the liquid slug.

The pressure of vapor and liquid are

$$p_1 = p_{0,v} + \frac{\pi}{2}\rho_v r_0 \omega^2 R, \ \omega \geq 0, \tag{3}$$

$$p_2 = p_{0,l} + \frac{\pi}{2}\rho_l r_0 \omega^2 R, \ \omega \geq 0, \tag{4}$$

where $p_{0,v}$ and $p_{0,l}$ are the pressure of the vapor and liquid phases under a static state, R is the distance between the OHP and the rotational axis and ω is the angular velocity.

The shear stress on the wall can be expressed as:

$$\tau_w = \mu\left(-\frac{du_l}{dr}\right)_w \tag{5}$$

According to the solution proposed by Ma [29], Equation (1) can be solved as follows

$$\rho_v r_0 u_{l,max}^2 \int_0^1 u_l^{*2} r^* dr^* = \frac{\pi}{4}\omega^2 r_0^3 R(\rho_l - \rho_v) + \frac{r_0}{2}(p_{0,l} - p_{0,v}) + \sigma(\cos\alpha_r + \cos\alpha_a) + \left(\frac{1}{r_0}\mu_l\left(-\frac{du^*}{dr^*}\right)_{r^*=1}\int_0^{L_l} u_{l,max} dx\right) \tag{6}$$

where $u_{l,max}$ is the maximum velocity of the liquid phase and μ_l is the dynamic viscosity of the liquid.
The following parameters are defined as follows:

$$A = \int_0^1 u_l^{*2} r^* dr^*, \tag{7}$$

$$B = \left(-\frac{du^*}{dr^*}\right)_{r^*=1}, \tag{8}$$

$$C = \frac{r_0}{2}(p_{0,l} - p_{0,v}) + \sigma(\cos\alpha_r + \cos\alpha_a) \qquad (9)$$

Therefore, Equation (6) will be rewritten as:

$$\rho_v r_0 u_{l,max}^2 A = \frac{\pi}{4}\omega^2 r_0^3 R(\rho_l - \rho_v) + C + \frac{B}{r_0}\mu_l \int_0^{L_l} u_{l,max} dx \qquad (10)$$

Under the static state, the driving force of vapor and the total resistant force to overcome are:

$$\rho_v r_0 u_{l,max}^2 A = C + \frac{B}{r_0}\mu_l \int_0^{L_l} u_{l,max} dx \qquad (11)$$

Meanwhile, under the axial-rotating condition, the driving force of vapor and the total resistant force to overcome is shown in Equation (10). It is clear that total resistant force is increased in case of an axial-rotating condition by $\pi\omega^2 r_0^3 R(\rho_l - \rho_v)/4$, so the vapor plug is more difficult to penetrate the liquid slug under axial rotation, and the annular flow will turn into slug flow under axial-rotation.

4. Results and Discussion

4.1. Effects of the Centrifugal Acceleration

The influence of rotation is described through the rotational speed and radius. Figure 6 illustrates the effects of centrifugal acceleration on the thermal performance of the axial-rotating single-loop oscillating heat pipe (SLOHP) filled with methanol, acetone and deionized water (DI water). It is clear that the centrifugal acceleration has a positive influence on the heat transport capacity of the axial-rotating SLOHP filled with methanol and acetone. When the centrifugal acceleration increases up to 30 m/s^2, the thermal resistance decreases rapidly for all the fluids. In the range of the centrifugal acceleration from 30 m/s^2 to 738 m/s^2, the thermal performance enhances for most cases of the axial-rotating SLOHP filled with methanol and acetone. In the range of 0 m/s^2 to 738 m/s^2, the thermal resistance decreases by 22% and 137% for the axial-rotating SLOHP filled with methanol and acetone, respectively. According to previous studies [19,25,28,30], the inside flow patterns can be estimated through the temperature curves. In the evaporator, the longer the vapor plug is, the longer it passes through the temperature measuring point, and the slower the vapor moves, the longer it passes through the measuring point. This leads to a higher amplitude of the temperature, as the hot long vapor plug will increase the temperature of the damped measuring point gradually until a liquid slug passes the measuring point (see Section 3.1). Moreover, the temperature differences decrease when the fluid circulation is faster and the vapor plugs and liquid slugs become shorter.

In order to emphasize the influence of the centrifugal acceleration on the flow pattern, a rotational speed of 950 rpm (296 m/s^2) is applied on the SLOHP filled with methanol under the heat flux of 27,300 W/m^2. Figure 7a shows the evaporator temperature curve. Under static conditions ($a = 0$ m/s^2), the value of the evaporator temperature is high, with the peak-to-valley value being around 20 °C, and the evaporator temperature fluctuating at low frequencies. On the contrary, under the centrifugal acceleration of 296 m/s^2, the temperature curve is obviously smoother, with a smaller amplitude and high oscillation frequencies. The peak-to-valley amplitude is around 10 °C under these conditions. In terms of the methanol and acetone, the centrifugal acceleration likely changes the annular flow to slug flow, as it is qualitatively proven by the theoretical analysis in Section 3.2, and increases the circular motion velocity. As a result, for methanol and acetone as working fluids, the thermal performance of the axial-rotating SLOHP enhances with the increase of centrifugal acceleration.

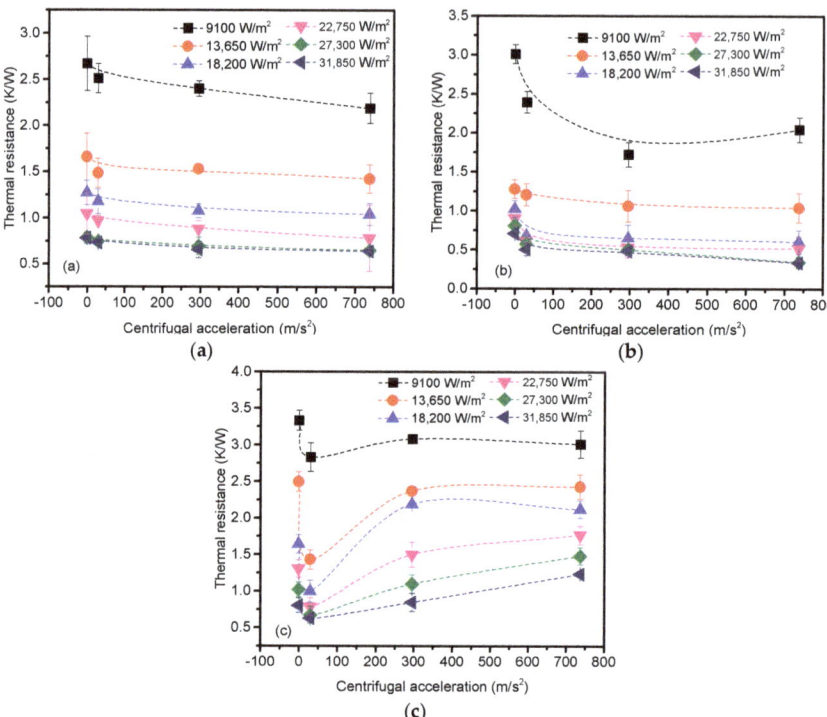

Figure 6. Effects of centrifugal acceleration: (**a**) methanol; (**b**) acetone; and (**c**) DI water.

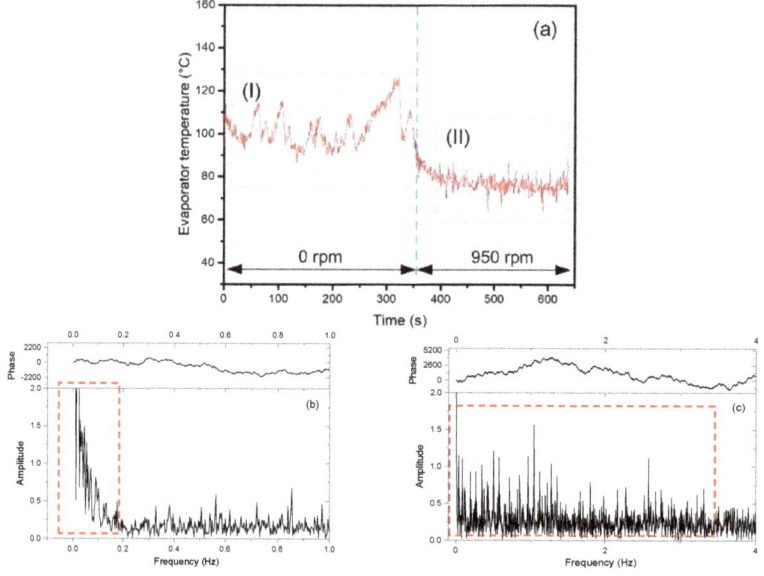

Figure 7. Evaporator temperature under rotational speed of 0 rpm and 950 rpm (filled with methanol): (**a**) temperature curve; and (**b**) and (**c**) FFT analysis of the temperature curve in areas (I) and (II).

On the contrary, as shown in Figure 6c, a different behavior is observed when DI water is the working fluid. Up to $a = 30$ m/s^2, the heat transfer performance increases sharply, the thermal resistance drops by up to 74%, from 2.5 K/W to 1.43 K/W, under the heat flux of 13,650 W/m^2. As the centrifugal acceleration exceeds 30 m/s^2, the thermal resistance rises gradually, which means the thermal performance deteriorates. Specifically, when the centrifugal acceleration is higher than 296 m/s^2, the thermal performance is lower than that of the static state at the heat flux range from 18,200 W/m^2 to 31,850 W/m^2. Therefore, in terms of the axial-rotating SLOHP filled with DI water, the thermal performance first enhances and then decreases along with the increase of centrifugal acceleration, which peaks at 30 m/s^2. This result is consistent with previous studies [22,25]. The reasons for such different behavior of water can be found in its higher viscosity (Table 2), which increases the flow friction [28], the higher thermal conductivity, which can produce local transient dry-out [25], and the lower value of $(dp/dT)_{sat}$, which brings it to a weaker expansion, i.e. a weaker flow circulation (see also Section 3.1). From the visualization, it is observed that under 300 rpm, the flow of water is a discontinuous train of vapor bubble and liquid slug. Meanwhile, when the speed exceeds 300 rpm, it is difficult to form effective circulation. An intermittent train of bubble and slug forms, and sometimes moves to the condenser. Most of the time, no obvious motion is observed—just local oscillation.

4.2. Effects of the Heat Flux

When the heat flux increases from 9100 W/m^2 to 31,850 W/m^2, the thermal performance of the axial-rotating SLOHP enhances at all centrifugal accelerations for all working fluids (Figure 8). In detail, for the axial-rotating SLOHP filled with methanol, the thermal resistance decreases from around 2.67 K/W to around 0.64 K/W by up to 4.5 times. As for the acetone as the working fluid, the thermal resistance decreases from around 3.01 K/W to around 0.34 K/W by a maximum of 8.6 times. Similarly, in terms of DI water, the thermal resistance decreases from around 3.33 K/W to around 0.63 K/W by four times. Based on our previous research [19], as the heat flux increased, changes of flow pattern and motion modes—which develops from bubbly flow to vapor plug flow, then to annular flow, and from oscillation to circulation—enhance the OHP's thermal performance.

4.3. Effects of the Working Fluid

From Table 2, acetone and methanol have a high $(dp/dT)_{sat}$ value, low dynamic viscosity and low surface tension, which lead to a higher driving force and the lower resistance of flow motion. On the other hand, DI water has a higher thermal conductivity, latent heat and specific heat. Due to differences in thermophysical properties, the working fluid has an important impact on thermal performance under the axial-rotating condition. The influence of the working fluids is different under low and high centrifugal accelerations, as illustrated in Figure 9. The axial-rotating SLOHP filled with acetone has the highest heat transport capacity, regardless of the heat flux and centrifugal acceleration. At the low centrifugal acceleration (e.g., 30 m/s^2), the effective heat transfer coefficient of the axial-rotating SLOHP filled with DI water is higher than that of methanol in general. However, when the centrifugal acceleration is high (e.g., 738 m/s^2), the thermal performance of the axial-rotating SLOHP filled with methanol is better than that of DI water. This is due to different flow patterns and motion modes, which are caused by the differences of thermophysical properties at varied centrifugal accelerations. In the light of this—for the given temperatures and heat load ranges, and even if the deeper reasons are not completely clear—as a conclusion, acetone is the recommended working fluid for this axial-rotating SLOHP, which is supposed to work in the abrasive-milling tool to enhance heat transfer in abrasive-milling processes.

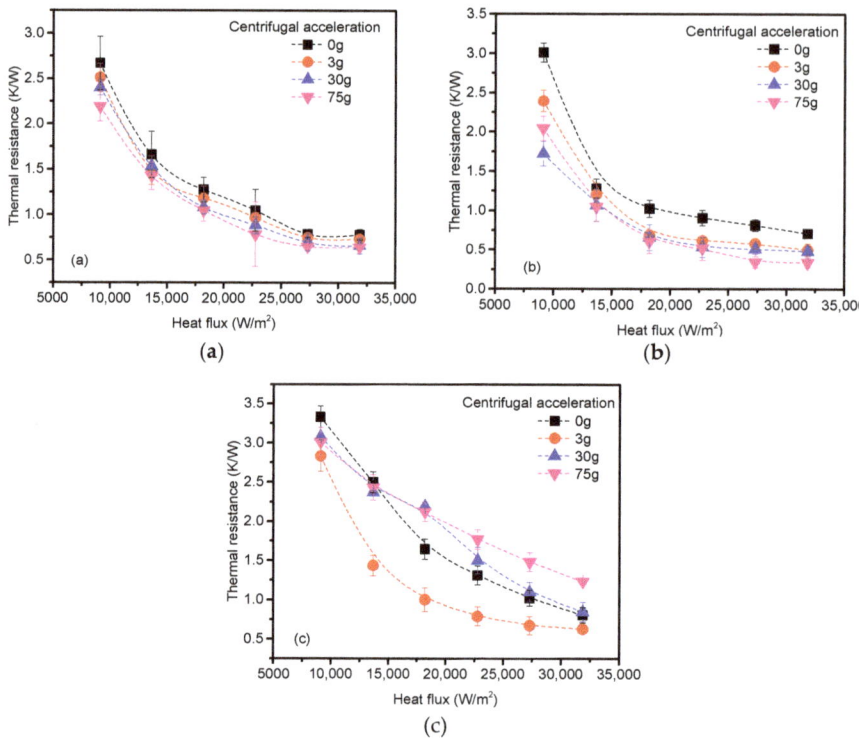

Figure 8. Effects of heat flux: (**a**) methanol; (**b**) acetone; and (**c**) DI water.

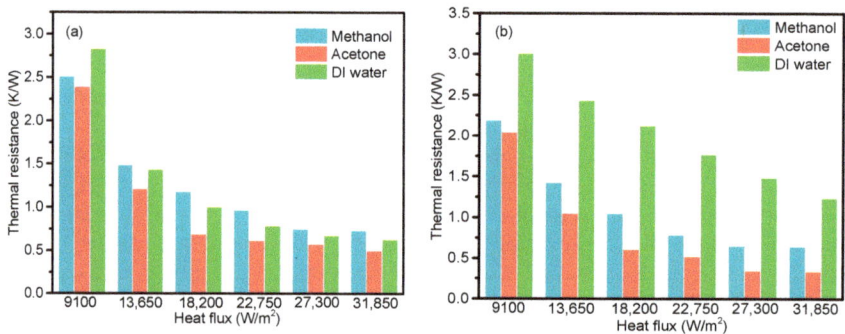

Figure 9. Effects of working fluids under centrifugal acceleration of: (**a**) 30 m/s^2; and (**b**) 738 m/s^2.

5. Conclusions

The influences of centrifugal acceleration, heat flux and working fluid on the thermal performance of the axial-rotating SLOHP are investigated through visualization, theoretical analysis and experiments. The main conclusions are drawn as follows:

The flow inside axial-rotating SLOHP is analyzed through the theoretical and visualization method. Based on theoretical analysis, the rotation will increase the resistance for the vapor to penetrate through the liquid slugs to form an annular flow, which is verified by the visualization. Besides, under the static state and heat flux of 22,750 W/m^2, the flow pattern of working fluid, acetone, is a veering circulation with the annular flow and slug flow. Meanwhile, under the axial-rotating conditions, the flow turns into slug flow. When the rotating speed increases from 60 rpm to 300 rpm, the flow changes from veering circulation to unidirectional circulation, and the liquid slug becomes shorter with a higher generating frequency.

For the acetone and methanol as working fluids, the thermal performance is enhanced by increasing the centrifugal acceleration, which is demonstrated through the change of internal flow motions. Nevertheless, the thermal performance of the axial-rotating SLOHP filled with DI water enhances first, and then decreases when the centrifugal acceleration increases. The acetone is recommended since the axial-rotating SLOHP filled with acetone has the best thermal performance. Furthermore, the heat transport capacity of the axial-rotating SLOHP improves with the increase of heat flux for all cases.

Author Contributions: Conceptualization, N.Q., Y.F. and J.X.; methodology and analysis, N.Q., M.M. and F.J.; writing—original draft preparation, N.Q., J.C.; writing—review and editing, Y.F., M.M. and J.X.; supervision, Y.F., M.M. and J.X.; funding acquisition, Y.F., M.M. and J.C. All authors have read and agreed to the published version of the manuscript.

Funding: The authors gratefully acknowledge the financial support for this work by the National Natural Science Foundation of China (Grant Nos. 51175254 and 51905275), National Natural Science Foundation of China for Creative Research Group (Grant No. 51921003) and Natural Science Foundation of Jiangsu Province (Grant No. BK20190752). Marco Marengo has been supported by UK EPSRC through the grant EP/P013112/1 (HyHP project, https://blogs.brighton.ac.uk/hyhp/).

Acknowledgments: The authors would like to address a great thank to Jiusheng Xu for his assistant to build the apparatus.

Conflicts of Interest: The authors declare no conflict of interest.

Nomenclature

a	Centrifugal acceleration, m/s^2
g	Gravity acceleration, 9.8 m/s^2
h_{eff}	Effective heat transfer coefficient, W/m^2·K
p	Pressure, Pa
Q	Heating power, W
q''	Heat flux, W/m^2
R	Thermal resistance, K/W (°C /W); distance from the rotational axis to the SLOHP, m
r_0	Radius of SLOHP tube, m
T	Temperature, °C
u	Velocity, m/s

Greek

α	Contact angle, °
μ	Dynamic viscosity, N·s/m^2
ρ	Density, kg/m^3
τ	Shear stress, N/m^2
ω	Angular velocity of rotation, rad/s

Subscripts

a	Advancing
cond	Condenser
eff	Effective
evap	Evaporator
l	Liquid
max	Maximum
sat	Saturated
r	Receding
v	Vapor
w	Wall
1	Vapor phase
2	Liquid phase

References

1. Qian, N.; Ding, W.; Zhu, Y. Comparative investigation on grindability of K4125 and Inconel718 nickel-based superalloys. *Int. J. Adv. Manuf. Technol.* **2018**, *97*, 1649–1661. [CrossRef]
2. Gavalda Diaz, O.; Garcia Luna, G.; Liao, Z.; Axinte, D. The new challenges of machining Ceramic Matrix Composites (CMCs): Review of surface integrity. *Int. J. Mach. Tools Manuf.* **2019**, *139*, 24–36. [CrossRef]
3. Liang, Y.; Chen, Y.; Chen, B.; Fan, B.; Yan, C.; Fu, Y. Feasibility of Ultrasonic Vibration Assisted Grinding for Carbon Fiber Reinforced Polymer with Monolayer Brazed Grinding Tools. *Int. J. Precis. Eng. Manuf.* **2019**, *20*, 1083–1094. [CrossRef]
4. Wang, Y.; Su, H.; Dai, J.; Yang, S. A novel finite element method for the wear analysis of cemented carbide tool during high speed cutting Ti6Al4V process. *Int. J. Adv. Manuf. Technol.* **2019**, 2795–2807. [CrossRef]
5. Jen, T.; Gutiérrez, G.; Eapen, S.; Barber, G.; Zhao, H.; Szuba, P.; Labataille, J.; Manjunathaiah, J. Investigation of heat pipe cooling in drilling applications. *Int. J. Mach. Tools Manuf.* **2002**, *42*, 643–652. [CrossRef]
6. Robinson Gnanadurai, R.; Varadarajan, A.S. Investigation on the effect of cooling of the tool using heat pipe during hard turning with minimal fluid application. *Eng. Sci. Technol. an Int. J.* **2016**, *19*, 1190–1198. [CrossRef]
7. Chiou, R.Y.; Chen, J.S.J.; Lu, L.; North, M.T. The Effect of an Embedded Heat Pipe in a Cutting Tool on Temperature and Wear. *Advances in Bioengineering* **2003**, 369–376.
8. He, Q.; Fu, Y.; Chen, J.; Zhang, W.; Cui, Z. Experimental investigation of cooling characteristics in wet grinding using heat pipe grinding wheel. *Int. J. Adv. Manuf. Technol.* **2018**, *97*, 621–627. [CrossRef]
9. He, Q.; Fu, Y.; Chen, J.; Zhang, W. Investigation on Heat Transfer Performance of Heat Pipe Grinding Wheel in Dry Grinding. *J. Manuf. Sci. Eng.* **2016**, *138*, 111009. [CrossRef]
10. Chen, J.; Fu, Y.; He, Q.; Shen, H.; Ching, C.Y.; Ewing, D. Environmentally friendly machining with a revolving heat pipe grinding wheel. *Appl. Therm. Eng.* **2016**, *107*, 719–727. [CrossRef]
11. Chen, J.; Fu, Y.; Gu, Z.; Shen, H.; He, Q. Study on heat transfer of a rotating heat pipe cooling system in dry abrasive-milling. *Appl. Therm. Eng.* **2017**, *115*, 736–743. [CrossRef]
12. Bastakoti, D.; Zhang, H.; Li, D.; Cai, W.; Li, F. An overview on the developing trend of pulsating heat pipe and its performance. *Appl. Therm. Eng.* **2018**, *141*, 305–332. [CrossRef]
13. Zhang, Y.; Faghri, A. Advances and unsolved issues in pulsating heat pipes. *Heat Transf. Eng.* **2008**, *29*, 20–44. [CrossRef]
14. Taft, B.S.; Williams, A.D.; Drolen, B.L. Review of Pulsating Heat Pipe Working Fluid Selection. *J. Thermophys. Heat Transf.* **2012**, *26*, 651–656. [CrossRef]
15. Alhuyi Nazari, M.; Ahmadi, M.H.; Ghasempour, R.; Shafii, M.B. How to improve the thermal performance of pulsating heat pipes: A review on working fluid. *Renew. Sustain. Energy Rev.* **2018**, *91*, 630–638. [CrossRef]
16. Stevens, K.A.; Smith, S.M.; Taft, B.S. Variation in oscillating heat pipe performance. *Appl. Therm. Eng.* **2019**, *149*, 987–995. [CrossRef]
17. Saha, N.; Das, P.K.; Sharma, P.K. Influence of process variables on the hydrodynamics and performance of a single loop pulsating heat pipe. *Int. J. Heat Mass Transf.* **2014**, *74*, 238–250. [CrossRef]
18. Mameli, M.; Manno, V.; Filippeschi, S.; Marengo, M. Thermal instability of a Closed Loop Pulsating Heat Pipe: Combined effect of orientation and filling ratio. *Exp. Therm. Fluid Sci.* **2014**, *59*, 222–229. [CrossRef]
19. Qian, N.; Fu, Y.; Zhang, Y.; Chen, J.; Xu, J. Experimental investigation of thermal performance of the oscillating heat pipe for the grinding wheel. *Int. J. Heat Mass Transf.* **2019**, *136*, 911–923. [CrossRef]
20. Khandekar, S.; Gautam, A.P.; Sharma, P.K. Multiple quasi-steady states in a closed loop pulsating heat pipe. *Int. J. Therm. Sci.* **2009**, *48*, 535–546. [CrossRef]
21. Wu, Z.; Deng, J.; Su, C.; Luo, C.; Xia, D. Performance of the micro-texture self-lubricating and pulsating heat pipe self-cooling tools in dry cutting process. *Int. J. Refract. Met. Hard Mater.* **2014**, *45*, 238–248. [CrossRef]
22. Wu, Z.; Yang, Y.; Luo, C. Design, fabrication and dry cutting performance of pulsating heat pipe self-cooling tools. *J. Clean. Prod.* **2016**, *124*, 276–282. [CrossRef]
23. Qian, N.; Fu, Y.; Marengo, M.; Chen, J.; Xu, J. Start-up timing behavior of single-loop oscillating heat pipes based on the second-order dynamic model. *Int. J. Heat Mass Transf.* **2020**, *147*, 118994. [CrossRef]
24. Qian, N.; Wang, X.; Fu, Y.; Zhao, Z.; Xu, J.; Chen, J. Predicting heat transfer of oscillating heat pipes for machining processes based on extreme gradient boosting algorithm. *Appl. Therm. Eng.* **2020**, *164*, 114521. [CrossRef]

25. Aboutalebi, M.; Nikravan Moghaddam, A.M.; Mohammadi, N.; Shafii, M.B. Experimental investigation on performance of a rotating closed loop pulsating heat pipe. *Int. Commun. Heat Mass Transf.* **2013**, *45*, 137–145. [CrossRef]
26. Ebrahimi Dehshali, M.; Nazari, M.A.; Shafii, M.B. Thermal performance of rotating closed-loop pulsating heat pipes: Experimental investigation and semi-empirical correlation. *Int. J. Therm. Sci.* **2018**, *123*, 14–26. [CrossRef]
27. Liou, T.-M.; Chang, S.W.; Cai, W.L.; Lan, I.-A. Thermal fluid characteristics of pulsating heat pipe in radially rotating thin pad. *Int. J. Heat Mass Transf.* **2019**, *131*, 273–290. [CrossRef]
28. On-ai, K.; Kammuang-lue, N.; Terdtoon, P.; Sakulchangsatjatai, P. Implied physical phenomena of rotating closed-loop pulsating heat pipe from working fluid temperature. *Appl. Therm. Eng.* **2019**, *148*, 1303–1309. [CrossRef]
29. Ma, H. *Oscillating heat pipes*; Springer: New York, NY, USA, 2015; p. 427. ISBN 9781493925049.
30. Monroe, J.G.; Aspin, Z.S.; Fairley, J.D.; Thompson, S.M. Analysis and comparison of internal and external temperature measurements of a tubular oscillating heat pipe. *Exp. Therm. Fluid Sci.* **2017**, *84*, 165–178. [CrossRef]

© 2020 by the authors. Licensee MDPI, Basel, Switzerland. This article is an open access article distributed under the terms and conditions of the Creative Commons Attribution (CC BY) license (http://creativecommons.org/licenses/by/4.0/).

Article

Experimental Validation of Heat Transport Modelling in Large Solar Thermal Plants

Kevin Sartor * and Rémi Dickes

Thermodynamics Laboratory, University of Liège, 4000 Liège, Belgium; rdickes@ulg.ac.be
* Correspondence: kevin.sartor@uliege.be

Received: 16 April 2020; Accepted: 6 May 2020; Published: 8 May 2020

Abstract: Solar thermal plants are often considered as a convenient and environmentally friendly way to supply heat to buildings or low temperature industrial processes. Some modelling techniques are required to assess the dynamic behaviour of solar thermal plants in order to control them correctly. This aspect is reinforced while large plants are considered. Indeed, some atmospheric conditions, such as local clouds, could have significant influence on the outlet temperature of the solar field. A common modelling approach to assess the heat transport in pipes is the one-dimensional finite volume method. However, previous work shows limitations in the assessment of the temperatures and in the computational time required for simulating large pipe networks. In this contribution, a previous alternative method developed and validated in a district heating network is used and extended to a solar thermal plant considering the thermal solar gain and the inertia of the pipes. The present contribution intends to experimentally validate this model on an existing solar plant facility available at the Plataforma Solar de Almeria in Spain.

Keywords: solar network; dynamic modelling; plug flow; control

1. Introduction

Solar thermal plants are one of the current solutions to change the world energy sources and to contribute to a green energy transition by recovering solar energy. The use of this kind of energy system has grown significantly for several years, especially to produce power or to supply heat to industrial processes or, in some cases, to residential buildings [1]. Moreover, among these systems, the concentration thermal plants tend to be economically competitive compared to conventional energy sources (60–100 EUR/MWh) [2], increasing their future use.

However, some modelling techniques are required to assess the dynamic behaviour of solar thermal plants to control them correctly. Indeed, like most renewable energy systems, a solar plant is facing to the intermittency of the energy source, namely the Sun, and requires some modelling techniques to couple them to other energy systems or energy storage facilities [3]. This aspect is reinforced while large plants are considered such as the solar park in Cape Town in South Africa, where the area of the solar plant can reach several km^2 [4]. In this plant configuration, some atmospheric conditions, such as local clouds, could have significant influence on the outlet temperature of the solar field and a control strategy should be used to maintain the outlet plant temperature near the defined set point as a predictive control based on climatic observations [5] or with dedicated algorithms [6]. Another solution is to instrument the network at numerous key locations to measure the operating conditions (temperatures and mass flow rates). However, this method is often expensive because of the numerous expensive sensors used.

A common modelling approach to assess the heat transport in pipes is the one-dimensional finite volume method. However, previous works [7–9] show limitations in the assessment of the temperatures and in the computational time required for simulating large pipe networks. Indeed, the finite volume method requires a high discretisation scheme to get accurate results. However, the increasing discretisation leads to a higher computational time. This computational time could be incompatible with the control strategy.

The present contribution intends to use an existing model developed by the authors dedicated to district heating networks and to extend it to assess the dynamic behaviour of solar thermal networks. Indeed, the proposed model, based on a Lagrangian approach [10], was validated considering the working fluid as water and a reduced heat transfer between the pipes and the ambient due to the large insulation of the pipes. In this contribution, the working fluid considered is oil and the heat transfer is higher due to the thermal power recovered from the Sun. Therefore, the model is extended into solar thermal plants, considering the thermal solar gain and the inertia of the piping system. The case study used to validate the dynamic model is a parabolic trough collectors (PTC)-based solar plant facility installed at the Plataforma Solar de Almeria in Spain.

The developed model allows for a further study to investigate control strategy definitions and their optimisation. Although this paper focuses on the experimental validation of a medium-size solar thermal plant, it could be extended to the modelling of larger solar plants, while the key element behaviour, namely a pipe, is studied and validated.

2. Problem Statement

A typical solar thermal plant is composed of a piping system dedicated to transport the energy recovered from the Sun to a defined process. In large solar thermal plants, the total pipe network length can reach over several kilometres. Therefore, the heat propagation through the network depends on the fluid velocity, which can induce some delay. These delays could reach some minutes or even hours, depending of the operating conditions and the pipe length. Indeed, the order of magnitude of the working fluid velocity is generally some meters per second to limit the pressures losses and the related electric pump consumption and it depends on the control strategies of the plant.

Previous works of the authors [7,11] show that a one-dimensional finite-volume method is able to model the dynamic behaviour of this kind of network. However, this modelling method requires an important spatial discretisation of the pipe to get accurate results. This spatial discretisation leads to high computational time requirements, which is not compatible with the final purpose of the modelling, i.e., the investigation and the optimisation of control strategies. For lower spatial discretisation layouts, a phenomenon named "numerical diffusion" occurs [7,8,12] and reduces the accuracy of the finite volume method by anticipating the heat waves at the pipe outlet.

Therefore, it is proposed to counter these issues by considering an alternative modelling method called the "plug flow" method, which is briefly detailed in the following section. The plug flow method accuracy is of the same order of magnitude as the one of the finite-volume method, with important spatial discretisation. Moreover, the simulation speed of the plug flow method is improved and is clearly one advantage for the definition and the optimisation of control strategies, such as model-based predictive control methods. Indeed, in large thermal networks, the weather conditions, and more specifically, the solar irradiance, can have an important influence on the performance of some parts of the plant, especially if a given outlet temperature is required for the supplying process connected to the solar thermal plant.

The plug flow modelling method has already been validated experimentally on a district heating network. However, such a district heating network involves a very low heat transfer between the pipe and the ambient due to the high insulation used on the piping system. In order to go a step further, this contribution proposes to experimentally validate the "plug flow" method on a solar thermal network facility characterised by a very high transfer. More specifically, the plug flow model is used to simulate a field of PTC, as described in Section 4.

3. Modelling

Regarding the flow inside the pipe, the proposed modelling method is derived on the standard TRNSYS Type 31 component. This modelling method is based on a Lagrangian approach, i.e., the properties of each fluid particle are considered along with their direction in function of time, considering the energy balance [10]. In order to simplify the resolution of the whole system, the flow is considered as incompressible, which is valid if the fluid is a liquid and for low pressure variations [13]. Moreover, the wall friction dissipation and the energy coming from the pressure drop into the pipe are neglected as the axial diffusion. This method was previously analysed, developed and validated under the Matlab platform for different operating conditions and pipe layouts [7,14], but has now been successfully ported to Modelica language [11] and the related open-source Modelica library is available in [15].

The current model is based on a simplified combination of energy and continuity equation (Equation (1)) [11]:

$$\frac{\partial(\rho c_p A T)}{\partial t} + \frac{\partial(\rho c_p A v T)}{\partial x} = -\dot{q}_e \tag{1}$$

where ρ denotes the density, c_p is the fluid specific heat, A is the pipe cross section (m^2), v is the flow velocity, x is the spatial coordinate, t is the time, T is the temperature and $\dot{q}_e$ is the heat loss per unit length (which is positive for heat loss from pipe to the ground).

Considering in the Lagrangian approach that the observer is attached to a fluid parcel, there is no notion of spatial coordinate. Finally, after rearranging the previous equation, it is possible determine the outlet pipe temperature (T_{outlet}) through the Equation (2):

$$T_{outlet} = T_{ground} + (T_{inlet} - T_{ground}) \cdot \exp\left(-\frac{\Delta T}{RC}\right), \tag{2}$$

ΔT is the delay time of the fluid parcel travelling the pipe; R is the thermal resistance between the fluid temperature and the ground temperature, which can be calculated by [16] or [17] depending on the geometric characteristics of the pipe, and C is the heat capacity per unit length of the water into the pipe. In this case, the fluid temperature is assumed uniform throughout the cross section of the pipe.

In the Modelica modelling, a unique volume is considered for the pipe. To assess the time delay of a fluid parcel between the inlet and the outlet of a pipe, the following dynamic equation (Equation (3)) is solved:

$$\frac{\partial z\,(y,t)}{\partial t} + v(t)\frac{\partial z\,(y,t)}{\partial y} = 0, \tag{3}$$

where $z\,(y,t)$ is the transported quantity, y is the normalised spatial coordinate, t is the time and $v(t)$ the normalised velocity. An approximation of the one-way wave equation was successfully introduced with the spatialDistribution() operator defined in the Modelica Language Specification [18]. This modelling method considers the heat propagation, the thermal inertia of the piping system, the pressure losses and the heat transfer with the ambient.

The hydraulic behaviour is assessed by a previously developed model denoted HydraulicDiameter of the Annex 60 Library [15]. In this case, the pressure drop (ΔP) is coupled to the mass flow rate ($\dot{m}$) using a quadratic relation as in (4):

$$\dot{m} = k\sqrt{\Delta P}, \tag{4}$$

where k is a constant depending on the piping system in function of the nominal conditions (nominal pressure losses for a nominal mass flow rate). A thermal capacity is added to the core pipe model at the outlet of the pipe to account for the thermal inertia of the pipe constituting material as [19]. The interested reader could refer to [11] for further details.

Regarding the heat transfer to the pipe, the radial heat balance between all the heat collection element components (see Figure 1) is modelled according to the deterministic model proposed by Forristal [20], which accounts for the most important phenomena, i.e.,:

- conduction and thermal energy accumulation in the metal pipe;
- convection and radiation between the glass envelope and the metal pipe;
- conduction and thermal energy accumulation in the glass envelope;
- radiation and convection losses to the environment.

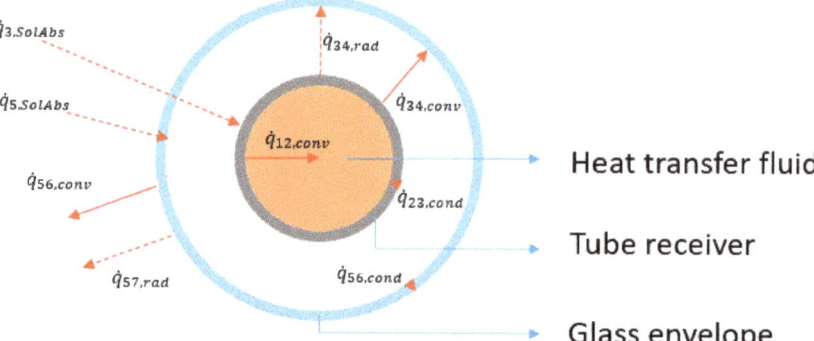

Figure 1. Energy balance around the heat collection element. In blue, the glass envelope; in grey, the metal pipe; and in white, the vacuum between the two. Heat transfer is highlighted with red arrows. Based on [21].

It is assumed that temperatures, heat transfer coefficients and thermodynamic properties are considered uniform around the circumference of the heat collector. Moreover, thermal losses through the support brackets are neglected and solar absorption in the tube and the glass envelope is treated as a linear phenomenon [21]. For further information about the original model, the reader can refer to [21–23].

For larger network where several pipes are interconnected, energy, mass and momentum balances could be performed on the key location to assess the energy and mass repartition inside the whole system according to [18].

4. Experimental Apparatus

The reference data used for the model validation are experimental measurements gathered on the Parabolic Trough Test Loop (PTTL) facility installed at the Plataforma Solar de Almería in Spain [24]. The data are recovered from a previous work presenting a finite-volume dynamic model of parabolic trough collectors [21]. As depicted in Figure 2, the facility includes three parallel lines of PTC from different manufacturers (AlbiasaTrough, EuroTrough and UrssaTrough) but only the EuroTrough collectors (ETC) are used in this work. The ETC line is composed of 6 EuroTrough modules connected in series and 18 prototype receiver tubes from a Chinese manufacturer, for a total length of 70.8 m and a net aperture area of 409.9 m^2. The system is a closed loop, with an East-West orientation and it is charged with the thermal oil Syltherm 800 [25].

The system operation is quite straightforward. A centrifugal pump drives the fluid from the point (1) through one of the three parallel PTC lines of the solar field. In the collectors (i.e., from points (2) to (3)), the heat transfer fluid is heated by absorbing solar energy reflected by the collectors onto the receiver tubes. The fluid is then cooled down by two air-cooled heat exchangers characterised by a maximum thermal capacity of 400 kWth. A 1 m^3 expansion vessel (filled by nitrogen) is placed in between the two air coolers to regulate the loop pressure (18 bar max). Finally, two additional electric heaters are installed at the outlet of the pump to control the oil temperature at the PTC inlet port.

The temperatures at the inlet and at the outlet of the PTC are measured with temperature transmittance sensors (denoted T in Figure 2). The direct normal irradiation (DNI) is measured with a pyrheliometer model CH1 by Kipp&Zonen [26]. A weather station installed nearby the solar field is used to measure the ambient conditions (temperature, humidity, wind speed and direction). The data acquisition system is based on National Instrument devices with a sampling time of 5 seconds. LabView software is used for data visualisation. The mass flow rate is derived from a data calibration study performed on the pump to assess it in function of the working conditions.

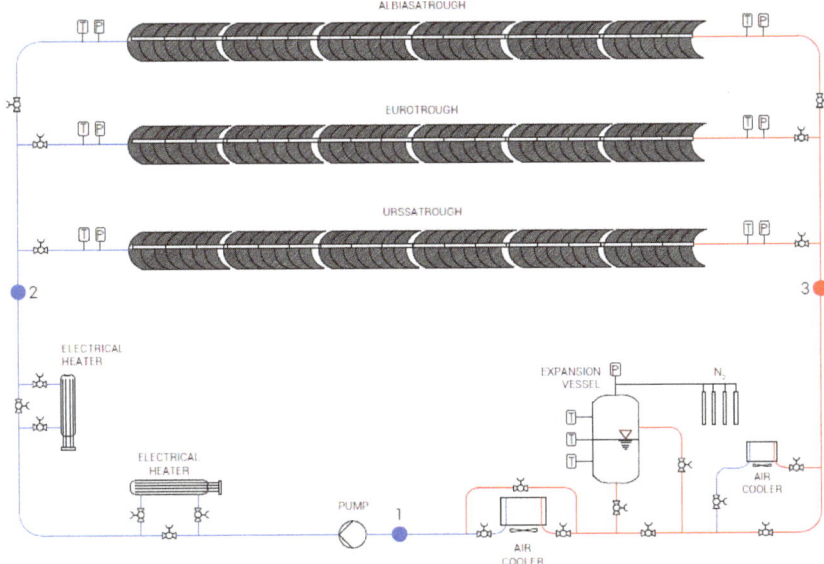

Figure 2. Process flow diagram of the Parabolic Trough Test Loop (PTTL) facilities with the relative sensors and devices position [21].

In Table 1, the working conditions of the main variables and of the external ambient parameters during the experimental campaign are reported.

Table 1. Range of the operation of the EuroTrough collectors (ETC) main variable and of the external ambient condition during the experimental campaign.

Variable	$\dot{m}$	T_{su}	T_{ex}	DNI	T_{amb}	v_{wind}
Unit	kg/s	°C	°C	W/m^2	°C	m/s
Min	1.55	150.05	170.21	593.95	26.23	0
Max	5.03	304.48	352.28	883.72	33.16	11.23

In order to perform the dynamic validation of the modelling method, three different scenarios are tested on the system:

- Oil mass flow change experiment (MFE): a step change is imposed onto the oil mass flow rate at the inlet of the ETC by varying the pump speed upwards or downwards, starting from a steady-state condition.
- Oil inlet temperature change experiment (TE): the oil temperature at the inlet of the ETC is varied by shutting down the air cooler, starting from a steady-state condition.
- Solar beam radiation change experiment (SBE): a step change to the solar beam radiation collected on the receiver is imposed downwards and upwards to the parabolic trough collectors by defocusing and focusing the parabolic trough collectors.

5. Results and Discussion

In this section, the model presented in Section 3 is validated onto the data presented in Section 4. The simulation is performed under Dymola2017 using Modelica language and the Differential Algebraic System Solver [27] with a relative tolerance of 10^{-4}. The results of [21], obtained with the one-volume finite modelling, are also used to compare the accuracy of the current model with this other common modelling method.

A good agreement is found between the experimental data and the modelling results as it can be seen in the figures available into this section. In Figure 3a,b, only TE and MFE scenarios are considered. When the mass flow rate increases (resp. decreases), the outlet pipe temperature decreases (resp. increases) a few times after, due to the quasi-constant energy transmitted to the pipe (during this experiment the DNI is quite constant). On the other hand, a heat wave propagation induced by an increase of the inlet temperature is also correctly modelled, while the trend of the outlet pipe temperature follows the experimental data.

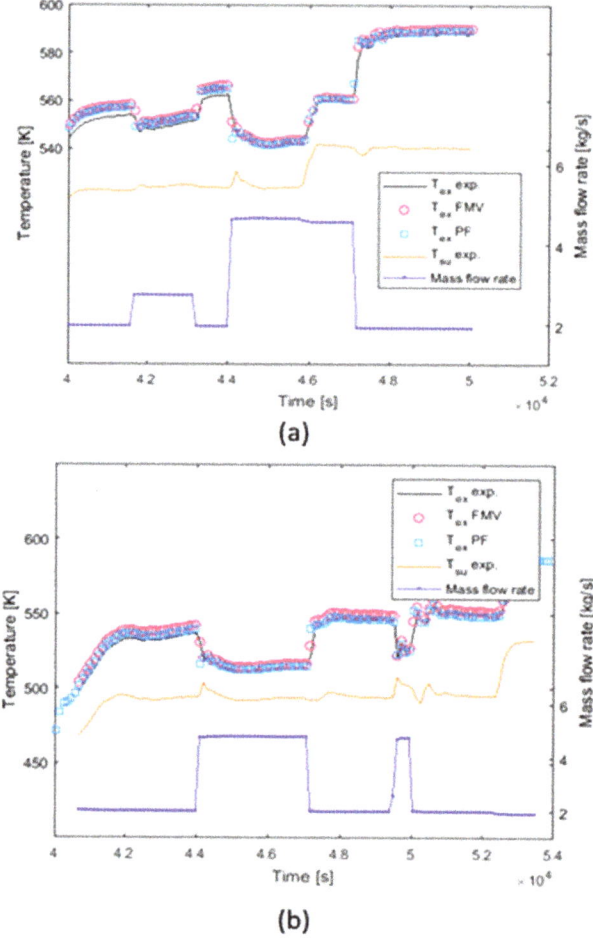

Figure 3. Experimental results performing different days versus plug flow and finite volume methods for several inlet temperature and mass flow rate conditions. (**a**) Test conditions #1; (**b**) Test conditions #2.

For the sake of clarity, in Figure 4a,b, only the SBE steps are noted on the experimental trends (denoted SBE) while TE and MFE conditions can be deduced from the figures. These SBE conditions are used to determine the influence of a cloud passing over the PTC system.

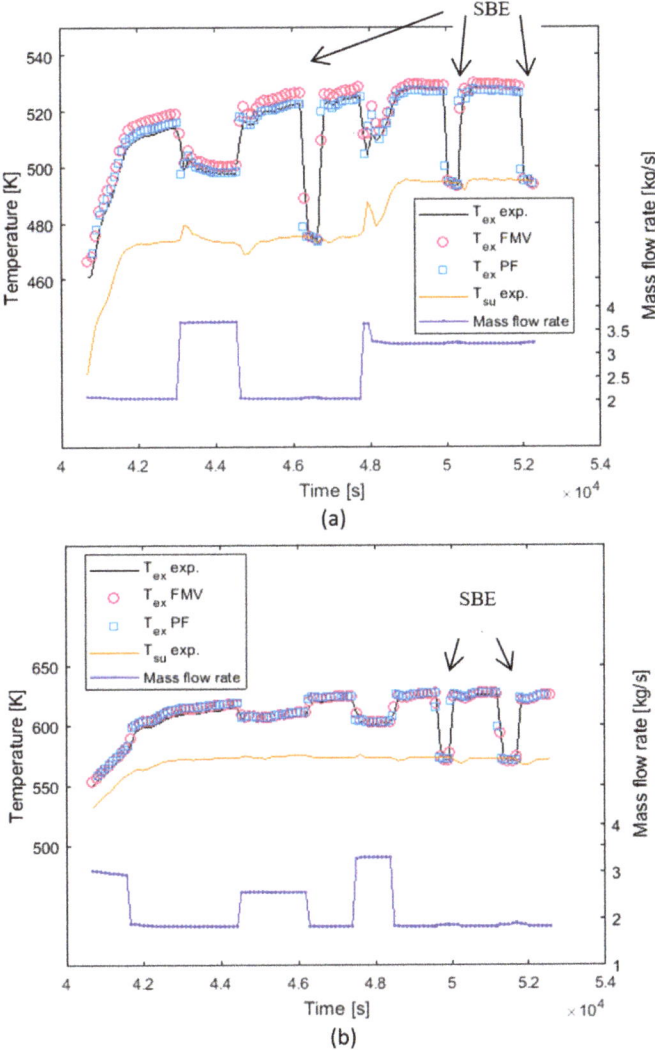

Figure 4. Experimental results performing different days versus plug flow and finite volume methods for several inlet temperature, mass flow rate and solar beam conditions. (**a**) Test conditions #3; (**b**) Test conditions #4.

Once again, the evolution of the predicted outlet temperature follows the experimental trends for several mass flow steps, inlet temperature steps and solar energy steps or a combination of them (Figure 4).

In all the cases studied, the accuracy is of the same order of magnitude as the finite volume method, with a discretisation of 50 cells which is the reference case of the study [21].

While the aim of this study is to consider active control strategies, the last experiment is presented in Figure 5 to assess the dynamic behaviour of the system when combinations of operating conditions steps are considered. Once again, there is a good agreement between plug flow modelling and the experimental data for wide variations of the mass flow rate, inlet pipe temperature and solar beam.

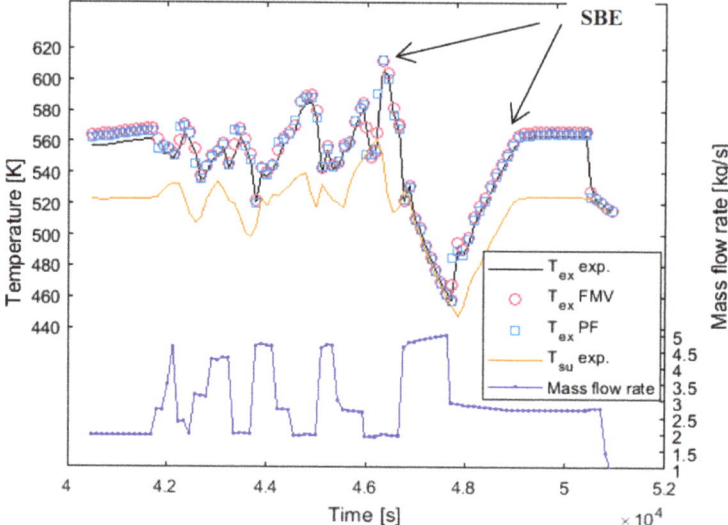

Figure 5. Experimental results performing one day versus plug flow and finite volume methods for a combination of inlet temperature, mass flow rate and solar beam step conditions.

As shown in previous study of district heating network modelling, the thermal inertia of the system and the operating mass flow rate have a significant influence on the outlet temperature response. Indeed, the mass flow rate influences how the fluid is propagated into the pipe, while the influence of thermal inertia can be seen when SBE conditions are set and can lead to significant extra delay (over 200 s) to get an outlet temperature with a similar value as the inlet temperature (considering the heat transfer to the ambient constant).

In this contribution, the simulation time required to model the process with plug flow modelling is generally reduced from a factor 20 to 30 compared to the finite volume method. This variation depends on the operating conditions and their trends. While this factor could seem quite reduced, it can be increased drastically in a larger network due to the higher complexity of the general problem. Indeed, in large networks with some intersections between the pipes, the hydraulic part of the problem can become complex to solve due to quadratic law relationship between pressure losses and the mass flow rates.

6. Conclusions

Solar thermal networks take place in the world energetic transition to supply clean thermal energy to process or residential needs. To optimise them and develop dedicated control strategies, it is required to assess the state of the heat transfer fluid correctly in several key points of the network, depending on the operating conditions. A practical solution could be to instrument these plants, but this solution is generally expensive and is not always easy to implement due to technical constraints. A second solution consists of modelling the dynamic behaviour of the whole system based on few input data.

In this contribution, the use of a previously developed plug flow approach for the district heating network modelling is extended through an experimental validation on a solar thermal network test bench. This validation step is required to check the assessment of the dynamic behaviour of the plant.

Indeed, the main difference between district heating networks and thermal networks is the magnitude of heat transfer between the piping system and the ambient. This experimental validation considers inlet temperature and mass flow rate variations, as it could be happening in some control strategies. On the other hand, the weather influence is simulated by a modified solar energy recovered to the pipes through a modified and controlled focussing and defocussing of the PTC system.

All the results show a good agreement between experimental data and simulation results for a wide range of operating conditions of the thermal plant. Moreover, the accuracy of the model is similar to those of the one-dimensional finite volume method with a reduced simulation time by a minimal factor of 20.

The next step of this work will consist of the definition and the optimisation of dedicated control strategies of the thermal plant to ensure the best control of the temperature required by the energy process.

Author Contributions: Conceptualization, K.S.; methodology, K.S.; software, K.S. and R.D.; validation, K.S.; formal analysis, K.S.; investigation, K.S.; resources, K.S. and R.D.; data curation, K.S. and R.D.; writing—original draft preparation, K.S.; writing—review and editing, K.S.; visualization, K.S. and R.D.; supervision, K.S.; project administration, K.S. and R.D. All authors have read and agreed to the published version of the manuscript.

Funding: This research received no external funding.

Conflicts of Interest: The authors declare no conflict of interest.

Nomenclature

$\dot{m}$	Mass flow rate, kg/s
$\dot{q}_e$	Heat loss per unit length W/m
c_p	Specific heat capacity, J/kg °C
ρ	Density, kg/m^3
A	Pipe cross section, m^2
DNI	Direct normal irradiation W/m^2
FMV	Finite Volume Method
HTF	Heat transfer fluid
MFE	mass flow change experiment
P	Pressure, bar
PF	Plug Flow
PTC	Parabolic Trough Collector
PTTL	Parabolic Trough Test Loop
SBE	Solar beam radiation change experiment
T	temperature, °C
TE	inlet temperature change experiment
t	time, s
v	velocity, m/s
x	spatial coordinate
y	normalised spatial coordinate
Subscripts and superscripts	
amb	ambient
conv	convective (heat transfer)
ex	exhaust
rad	radiative (heat transfer)
SolAbs	Solar power absorbed
su	supply

References

1. Duffie, J.A.A.; Beckman, W.A. *Solar Engineering of Thermal Processes*; John Wiley & Sons, Inc.: Hoboken, NJ, USA, 2013.
2. Andrei, I.; Ralon, P.; Rodriguez, A.; Taylor, M.; International Renewable Energy Agency. Abu Habi: 2017. Available online: https://www.irena.org/-/media/Files/IRENA/Agency/Publication/2018/Jan/IRENA_2017_Power_Costs_2018.pdf (accessed on 7 April 2020).
3. Pramanik, S.; Ravikrishna, R.V. A review of concentrated solar power hybrid technologies. *Appl. Therm. Eng.* **2017**, *127*, 602–637. [CrossRef]
4. Mathews, J.A.; Hu, M.-C.; Wu, C.-Y. Concentrating solar power: A renewable energy frontier. *Carbon Manag.* **2014**, *5*, 293–308. [CrossRef]
5. Martínez, D.; Rubio, F.R.; Berenguel, M.; Camacho, E.F. *Control of Solar Energy Systems*; Springer-Verlag: London, UK, 2012. [CrossRef]
6. Schlipf, D.; Schneider, G.; Maier, H. Using evolutionary algorithm to develop a feed forward control for CSP plant using mid- and long term storages. *Energy Procedia* **2013**, *49*, 2191–2200. [CrossRef]
7. Sartor, K.; Thomas, D.; Dewallef, P. A comparative study for simulating heat transport in large district heating networks. *Int. J. Heat Technol.* **2018**, *36*, 301–308. [CrossRef]
8. Van den Bossche, G. Lokale Temperatuurverhoging Versus Tijdmodulatie in Lage Temperatuur Warmtenetten. Ph.D. Thesis, KULeuven, Leuven, Belgium, 2015.
9. Grosswindhager, S.; Voigt, A.; Kozek, M. Linear finite-difference schemes for energy transport in district heating networks. In Proceedings of the 2nd International Conference on Computer Modeling and Simulation, Brno, Czech Republic, 5–7 September 2011; pp. 5–7.
10. Bennett, A. *Lagrangian Fluid Dynamics*; Cambridge University Press: Cambridge, UK, 2006.
11. van der Heijde, B.; Fuchs, M.; Tugores, C.R.; Schweiger, G.; Sartor, K.; Basciotti, D.; Müller, D.; Nytsch-Geusen, C.; Wetter, M.; Helsen, L. Dynamic equation-based thermo-hydraulic pipe model for district heating and cooling systems. *Energy Convers. Manag.* **2017**, *151*, 158–169. [CrossRef]
12. Sartor, K. *Annex 60: Subtask 2.2 Modeling Heat Transport in District Heating Networks*; Annex 60: Paris, France, 2015.
13. Hoffman, J.; Johnson, C. *Computational Turbulent Incompressible Flow*; Springer: Berlin/Heidelberg, Germany, 2007; Volume 4. [CrossRef]
14. Sartor, K.; Dewalef, P. Experimental validation of heat transport modelling in district heating network. *Energy* **2017**, *137*, 961–968. [CrossRef]
15. Wetter, M.; Fuchs, M.; Grozman, P.; Helsen, L.; Jorissen, F.; Lauster, M. IEA EBC ANNEX 60 Modelica library—An international collaboration to develop a free opensource model library for buildings and community energy systems. In Proceedings of the BS2015 14th International Building Performance Simulation Association, Rome, Italy, 2–4 September 2019; pp. 395–402.
16. Bøhm, B. On transient heat losses from buried district heating pipes. *Int. J. Energy Res.* **2000**, *24*, 1311–1334. [CrossRef]
17. Velut, S.; Tummescheit, H. Implementation of a transmission line model for fast simulation of fluid flow dynamics. In Proceedings of the 8th International Modelica Conference, Dresden, Germany, 20–22 March 2011; p. 8.
18. Association, M. Modelica®—A Unified Object-Oriented Language for Systems Modeling Language Specification. 2014. Available online: https://www.modelica.org/documents/ModelicaSpec33Revision1.pdf (accessed on 7 May 2020).
19. Benonysson, A.; Bøhm, B.; Ravn, H.F. Operational optimization in a district heating system. *Energy Convers. Manag.* **1995**, *36*, 297–314. [CrossRef]
20. National Renewable Energy Laboratory. *Heat Transfer Analysis and Modeling of a Parabolic Trough Solar Receiver Implemented in Engineering Equation Solver*; National Renewable Energy Laboratory: Golden, CO, USA, 2003.
21. Desideri, A.; Dickes, R.; Bonillab, J.; Valenzuela, L.; Quoilin, S.; Lemort, V. Steady-state and dynamic validation of a parabolic through collector model using the ThermoCycle Modelica library. *Sol. Energy* **2018**, *174*, 866–877. [CrossRef]
22. Dahm, J. District Heating Pipelines in the Ground—Simulation Model. 2001. Available online: https://trnsys.de/download/de/ts_type_313_de.pdf (accessed on 26 September 2016).

23. TRNSYS 17 Manual—Volume 4—Mathematical Reference 2009, 486. Available online: https://docplayer.net/13373943-Trnsys-17-volume-4-mathematical-reference-a-transient-system-s-imulation-program.html (accessed on 23 December 2018).
24. León, J.; Clavero, J.; Valenzuela, L.; Zarza, E.; García, G. PTTL—A Life-size Test Loop for Parabolic Trough Collectors. In Proceedings of the SolarPACES 2013 International Conference, Las Vegas, NV, USA, 17–20 September 2013; Volume 49, pp. 136–144. [CrossRef]
25. Dow Inc. *Syltherm 800 Heat Transfer Liquid*; Dow Inc.: Midland, MI, USA, 1997.
26. Kipp & Zonen B.V. *CH1 Normal Incidence Pyrheliometer Manual*; Kipp & Zonen B.V.: Delft, The Netherlands, 1997.
27. Petzold, L.R. A Description of DASSL: A Differential-algebraic system solver. In Proceedings of the 10th IMACS World Congress, Montreal, QC, Canada, 8–13 August 1982; pp. 65–68.

© 2020 by the authors. Licensee MDPI, Basel, Switzerland. This article is an open access article distributed under the terms and conditions of the Creative Commons Attribution (CC BY) license (http://creativecommons.org/licenses/by/4.0/).

Article

Design Evaluation for a Finned-Tube CO_2 Gas Cooler in Residential Applications

Charalampos Alexopoulos [1], Osama Aljolani [2], Florian Heberle [2,*], Tryfon C. Roumpedakis [1], Dieter Brüggemann [2] and Sotirios Karellas [1]

1. Laboratory of Steam Boilers and Thermal Plants, Department of Thermal Engineering, School of Mechanical Engineering, National Technical University of Athens, 9 Heroon Polytechniou Street, 15780 Zografou, Greece; babisalexop@hotmail.com (C.A.); troumpedak@central.ntua.gr (T.C.R.); sotokar@mail.ntua.gr (S.K.)
2. Chair of Engineering Thermodynamics and Transport Processes (LTTT), Center of Energy Technology (ZET), Faculty of Engineering Science, University of Bayreuth, Universitätsstraße 30, 95440 Bayreuth, Germany; Osama.Aljolani@uni-bayreuth.de (O.A.); Dieter.Brueggemann@uni-bayreuth.de (D.B.)
* Correspondence: florian.heberle@uni-bayreuth.de

Received: 22 April 2020; Accepted: 10 May 2020; Published: 12 May 2020

Abstract: Towards the introduction of environmentally friendlier refrigerants, CO_2 cycles have gained significant attention in cooling and air conditioning systems in recent years. In this context, a design procedure for an air finned-tube CO_2 gas cooler is developed. The analysis aims to evaluate the gas cooler design incorporated into a CO_2 air conditioning system for residential applications. Therefore, a simulation model of the gas cooler is developed and validated experimentally by comparing its overall heat transfer coefficient. Based on the model, the evaluation of different numbers of rows, lengths, and diameters of tubes, as well as different ambient temperatures, are conducted, identifying the most suitable design in terms of pressure losses and required heat exchange area for selected operational conditions. The comparison between the model and the experimental results showed a satisfactory convergence for fan frequencies from 50 to 80 Hz. The absolute average deviations of the overall heat transfer coefficient for fan frequencies from 60 to 80 Hz were approximately 10%. With respect to the gas cooler design, a compromise between the bundle area and the refrigerant pressure drop was necessary, resulting in a 2.11 m^2 bundle area and 0.23 bar refrigerant pressure drop. In addition, the analysis of the gas cooler's performance in different ambient temperatures showed that the defined heat exchanger operates properly, compared to other potential gas cooler designs.

Keywords: supercritical carbon dioxide; experimental testing; finned-tube gas cooler

1. Introduction

The use of air conditioning systems is expanding rapidly around the world. An estimated amount of 700 million air conditioners will be operating in the world by 2030 [1]. This growing demand for air conditioning systems has enormous impacts on the environment.

Currently, a number of present regulations have been applied worldwide to control the use of harmful refrigerants [2–4]. The key implications of the use of conventional refrigerants include the depletion of the ozone layer and global warming. Based on the Montreal Protocol, a complete abolishment of chlorofluorocarbons (CFCs) was decided, due to their high ozone depletion potential (ODP) [5]. In addition, the phase out of hydrochlorofluorocarbon (HCFC) refrigerants was implemented [6]. On the other hand, the F-gas Regulation, first issued in the European Union in 2006, aimed to introduce measures for the reduction of fluorinated gases—hydrofluorocarbons (HFCs) and perfluorocarbons (PFCs)—in form of a phase-down, due to their high global warming potential (GWP) [7].

Instead, natural refrigerants are proposed as the substitute for the harmful refrigerants. Carbon dioxide (CO_2) is a natural, low cost, non-flammable, non-toxic refrigerant. Subsequently, it has emerged as a credible natural refrigerant to replace HFCs and HCFCs. However, its unique critical point, high critical pressure of 73.8 bar, and low critical temperature of 30.98 °C, remarkably affects the performance of CO_2 refrigeration systems, as well as imposes special design and control challenges. When the ambient temperature is higher than the critical temperature of CO_2, the system operates at supercritical conditions, and the heat rejection process occurs at a supercritical regime. In consequence, a phase change does not take place, and the heat exchanger in which this change of state occurs is called the gas cooler.

The impact of the gas cooler on the CO_2 refrigeration systems plays an important role, due to its high exergy loss. Therefore, it is considered vital to be further investigated and designed properly [8]. The finned-tube type for gas coolers is well established in the heating, ventilation, and air conditioning (HVAC) and refrigeration industries, due to its compactness and manufacturing flexibility. The design of the finned-tube heat exchangers affects considerably the overall heat transfer performance and system efficiency. Particularly, fin and tube thickness and the respective materials, spacing, and dimensions of the tubes and fins are crucial parameters of the design [9]. Fundamental studies about the heat transfer characteristics during the heat rejection process in tubes have been performed theoretically and experimentally by many researchers since Lorentzen and Pettersen [10] proposed the transcritical CO_2 cycle for mobile air conditioning systems.

Pitla et al. [11] conducted an investigation about heat transfer phenomena and pressure losses of CO_2 at supercritical conditions into a tube. They found that the majority of the deviations between the numerical and experimental values are within ± 20%, and a new heat transfer correlation was presented. Son and Park [12] carried out an experiment in order to investigate the gas cooling process of CO_2 in terms of heat transfer coefficient and pressure drop of the refrigerant. They described the variations of local heat transfer coefficient in the cooling process in the direction of the flow and proposed a more accurate heat transfer correlation. Zhang et al. [13] evaluated the performance of a printed circuit heat exchanger for cooling CO_2 with water. The analysis concluded that rapid variations in the thermodynamic properties of supercritical CO_2 increase entropy generation and therefore, to optimize the second law efficiency of the investigated heat exchanger, higher CO_2 mass flow rates should be used. Jadhav et al. [14] evaluated, using simulations, CO_2 gas coolers for air conditioning applications. For their investigation, a counter crossflow plain fin and staggered tube configurations were considered. According to the study, transverse tube spacing, gas cooler width, and air volumetric flow were the most influential parameters in the heat transfer mechanisms of the gas cooler.

Liu et al. [15] investigated experimentally the supercritical CO_2 characteristics in horizontal tubes with inner diameter of 4, 6, and 10.7 mm in terms of heat transfer phenomena and pressure losses. The authors concluded that the tube diameter significantly affects the heat transfer performance, and they proposed a new heat transfer correlation for the large diameter. Jiang et al. [16] investigated the convection heat transfer of CO_2 at supercritical pressures in a vertical small tube with inner diameter of 2.0 mm, experimentally and numerically. They studied the effects of various operational parameters and buoyancy on convection heat transfer in a small diameter. They concluded that when the CO_2 bulk temperatures are in the near-critical region, the local heat transfer coefficients vary significantly along the tube. Chai et al. [17] investigated, using simulations, the performance of finned-tube supercritical CO_2 gas coolers, combining a distributed modeling approach with the ε-NTU method. The results indicated that the performance of the gas coolers was enhanced by higher mass flow rates and lower tube diameters at the expense of higher pressure drops.

Other researchers have also investigated the performance of the air-cooled CO_2 gas coolers. Cheng et al. [18] presented an analysis of heat transfer and pressure drop experimental data and correlations for supercritical CO_2 cooling in macro- and micro-channels. Ge and Cropper [19] presented a detailed mathematical model for air-cooled finned-tube CO_2 gas coolers. They used a distributed method in order to obtain more accurate refrigerant thermophysical properties and local heat transfer coefficients during cooling processes. The model was compared with published test results. The comparison showed that the approach temperature and the heat capacity are simultaneously improved with the increase of heat exchanger circuit numbers. Marcinichen et al. [20] conducted simulations to optimize the working fluid charge of the gas cooler. The optimal design of the study reduced CO_2 charge by 14%, compared to a reference design. Moreover, the analysis revealed the importance of the oil concentrations in the CO_2 pressure drop, which is up to 2.65 times higher for oil concentrations of up to 3%. Zilio et al. [21] experimentally evaluated two different gas coolers, one with continuous, and one with separated fins, and on two different circuit arrangements for a transcritical CO_2 cycle. Using a coil with fins, a heat flux improvement of up to 5.6% was identified, which corresponded to a coefficient of performance (COP) increase of up to 6.6% for a conventional CO_2 refrigeration cycle.

Gupta and Dasgupta [22] applied a similar modelling method to the one from Ge and Cropper, [19] in order to evaluate the performance of the heat exchanger being affected by the airflow velocity. Here, a higher gas cooler performance is achieved at a higher air flow velocity as it decreases the refrigerant's approach temperature, and thus the heating capacity of the gas cooler is increased. Santosa et al. [23] built two CO_2 finned-tube gas coolers with different structural designs and controls, connected with a test rig of a CO_2 booster refrigeration system. They carried out experiments at different operating conditions while they developed models of the finned-tube CO_2 gas cooler. The analysis was conducted based on the distributed and lumped methods. They concluded that the heat exchanger design can affect the performance of both the component and the integrated system.

Although the heat transfer and pressure drop characteristics of the supercritical CO_2 in tubes have been investigated extensively using experimental and theoretical methods, research on the air-cooled finned-tube CO_2 gas coolers is still limited.

In this paper, mathematical calculations of the finned-tube CO_2 gas cooler are conducted, in order to establish a reliable design procedure. With focus on the heat transfer characteristics of the air side, the developed model is validated with an experimental setup using water as working fluid. Investigations of the effects of fan frequency, water inlet temperature, and water mass flow on the overall heat transfer coefficient are conducted, while deviations between the model and the test results are extracted according to the fan frequency. In addition, potential heat transfer correlations for the air- and refrigerant-side heat transfer coefficients have been studied. Finally, the model was applied to identify a reliable and efficient finned-tube CO_2 gas cooler design, as well as to evaluate its performance in different off-design conditions under varying ambient temperatures.

2. Materials and Methods

This study is part of a larger project of CO_2 air conditioning systems for residential applications and focuses on the gas cooler. A scheme of the considered CO_2 air conditioning system is depicted in Figure 1. Particularly, an efficient and reliable air finned-tube gas cooler was designed based on the boundary conditions, which are given in Table 1.

Table 1. On-design specifications of the gas cooler.

Property	Value
CO_2 inlet pressure (bar)	93
CO_2 temperature inlet/outlet (K)	358.22/311.15
CO_2 mass flow rate (kg s$_{-1}$)	0.146
Air temperature inlet/outlet (K)	308.15/315.15
Air mass flow rate (kg s$_{-1}$)	3.601
Heat duty (kW)	25.4

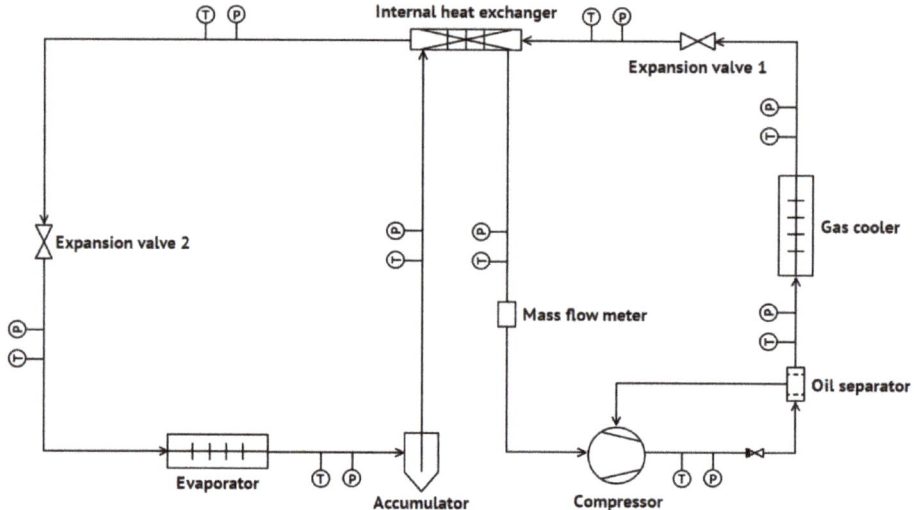

Figure 1. Scheme of the considered CO_2 air conditioning system.

In order to design the CO_2 gas cooler, a script in MATLAB R2019a [24] was developed. The simulation model calculated the overall heat transfer coefficient U of the gas cooler, based on the mass flows, inlet and outlet temperatures, and pressures of the medium. Subsequently, the required exchange area A_R was determined. Both parameters were based on the following equations:

$$U = \left[\frac{1}{h_{air}} + \frac{A_o \times \ln\frac{d_o}{d_i}}{2 \times \pi \times L \times k} + \frac{A_o}{A_i} \times \frac{1}{h_{refri}} \right]^{-1} \quad (1)$$

where A_o and A_i represent the outer and inner surface of the tube, respectively.

$$A_R = \frac{Q}{U \times \Delta T_{LMTD}} \quad (2)$$

The U-value consists of three parts: air convection, refrigerant convection, and the conduction. Compared to the other two heat transfer contributions, the conduction plays a minor role. The heat transfer coefficients on the air- and refrigerant-side are crucial for the overall heat transfer coefficient, and the validation for them are considered necessary. In order to validate the calculations, the overall heat transfer coefficient of a defined air-cooled heat exchanger was investigated experimentally.

The experimental part was based on a test rig using water and employed a specific design of an air-cooled heat exchanger (HEX). The HEX type was a finned tube with a fan air cooling system. The U-value was investigated experimentally for the entire HEX for different conditions. The second part of the validation consisted of modelling the heat exchanger to simulate the finned tube of the HEX.

2.1. Theoretical Model

For the model calculations, a script was created in MATLAB R2019a [24], modelling the defined gas cooler. The model overall heat transfer coefficient of the HEX was calculated from Equation (1).

2.1.1. Air-Side Heat Transfer

In-line arrangement and circular finned tubes were assumed, in order to model the HEX. Based on the assumption of crossflow type, the air- and refrigerant-side heat transfer coefficients can be calculated. The proposed correlation from VDI-Heat Atlas [25] was used in order to calculate the Nusselt number, using the following equation:

$$Nu = C \times Re_{d_o}^{0.6} \left(\frac{A_o}{A_{to}}\right)^{-0.15} Pr^{\frac{1}{3}} \qquad (3)$$

with C = 0.22 for in-line arrangement. A_o/A_{to} is the ratio of the finned surface to the surface of the base tube, and for circular fins was calculated from the following equation:

$$\frac{A_o}{A_{to}} = 1 + 2 \times \frac{H_f \times (H_f + d_o + t_f)}{s \times d_o} \qquad (4)$$

The Reynolds number was calculated by the equation:

$$Re_{d_o} = \frac{\rho_{air} \times w_s \times d_o}{\mu}, \qquad (5)$$

where w_s is the velocity in the smallest cross-section and was calculated from the following equation:

$$w_s = w_{inf} \times \frac{A_{inf}}{A_s} \qquad (6)$$

The air-side heat transfer coefficient was calculated from its definition, as the following equation shows.

$$h_{air} = \frac{Nu \times k_{air}}{d_o} \qquad (7)$$

However, the air-side heat transfer coefficient was affected by the fins. The fins should be taken into consideration, thus the following equation was used [25]:

$$h_{air,f} = h_{air} \times \left[1 - (1 - \eta_f) \times \frac{A_f}{A_o}\right] \qquad (8)$$

The fin efficiency is defined as the ratio of the heat removed by the fin to the heat removed by the fin at wall temperature. The efficiency of the fin was calculated from the following equation:

$$\eta_{fin} = \frac{\tanh X}{X}, \qquad (9)$$

with [25]:

$$X = \varphi \times \frac{d_o}{2} \times \sqrt{\frac{2 \times h_{air}}{k \times t_f}} \qquad (10)$$

and

$$\varphi = \left(\frac{d_f}{d_o} - 1\right)\left[1 + 0.35 \ln\left(\frac{d_f}{d_o}\right)\right] \qquad (11)$$

for circular fins [25]. So, the equation of the overall heat transfer coefficient used the updated air-side heat transfer coefficient as follows:

$$U = \left[\frac{1}{h_{air,f}} + \frac{A_o \times \ln \frac{d_o}{d_i}}{2 \times \pi \times L \times k} + \frac{A_o}{A_i} \times \frac{1}{h_{refri}}\right]^{-1} \qquad (12)$$

2.1.2. Refrigerant-Side Heat Transfer

On the refrigerant-side, the Gnielinski correlation [26] was used to calculate the Nusselt number:

$$Nu = \frac{\frac{f}{8}(Re_b - 1000)Pr}{12.7\sqrt{\frac{f}{8}}\left(Pr^{\frac{2}{3}} - 1\right) + 1.07} \tag{13}$$

which is valid in the range of $2300 < Re_b < 10^6$.

In addition, the friction factor f was calculated by:

$$f = [0.79 \ln(Re_b) - 1.64]^{-2} \tag{14}$$

Here, the Reynolds number was defined as

$$Re_b = \frac{G_{refri} \times d_i}{\mu} \tag{15}$$

and G was defined as the mass velocity, and was calculated from the following equation:

$$G_{refri} = \frac{m_{refri}/N_t}{\pi \times d_i^2/4} \tag{16}$$

Finally, heat transfer coefficient at the refrigerant side was calculated from:

$$h_{refri} = \frac{Nu \times k_{refri}}{d_o} \tag{17}$$

Further investigation of potential heat transfer correlations was conducted. More specifically, comparisons between the correlations of Gnielinski [26] and Dittus–Boelter [27], and the correlations proposed by VDI-Heat Atlas [25] and by Schmidt [28] were made for the refrigerant- and air-side, respectively. The equations below illustrate the Dittus–Boelter's [27] and Schmidt's [28] correlations, respectively:

$$Nu = 0.023 \times Re^{4/5} Pr^n \tag{18}$$

where $n = 0.3$ for the fluid being cooled.

$$Nu = C \times Re^{0.625} Pr^{1/3}\left(\frac{A_o}{A_{to}}\right)^{-0.375} \tag{19}$$

where $C = 0.3$ for in-line arrangement.

2.2. Experimental Set Up

As it is referred, the U-value consists of three parts: the air convection, refrigerant convection, and the conduction. In the present case, the air-side heat transfer represents the main thermal resistance. Thus, the experimental set up aimed to identify a suitable heat transfer correlation with focus on the air side. In order to validate the calculations, especially the air-side heat transfer, a well-known working medium should be used for the experiments on the refrigerant-side. Here, water with well-known thermophysical properties and reliable heat transfer correlations at single-phase regime was selected as a working medium.

The test rig consisted of the heater, the gas cooler, the measurement equipment, and controls. To enable the information to be read and recorded, the instrumentations were connected to a data logging system. The test rig is shown in Figure 2 below.

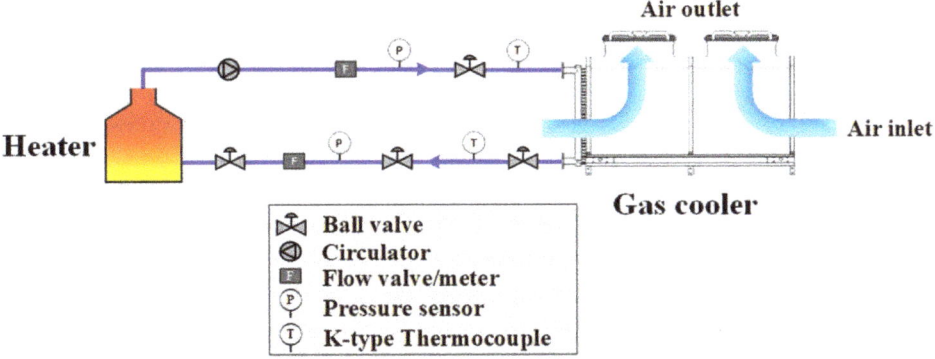

Figure 2. Schematic of the gas cooler test rig.

The air-cooled HEX used for the experiments was the CU-713CX2 from Panasonic. The heater was from the Single® company, model STW 150/1-18-45-KS7. The K-type thermocouples and pressure transducers used were from OMEGA company with uncertainties of ±0.2 °C and ±1.5%, respectively, while the mass flow valve and meter with uncertainties of ±0.5% were manufactured by Bürkert.

The experiments were carried out for different water mass flows, fan frequencies, and inlet water temperatures, as the following Table 2 shows. Figure 3a,b illustrate the air mass flow rate and fan power as function of the fan frequency.

Table 2. Range and interval of the experimental variables.

Range—Water Mass Flow Rate (L/min)	Range—Inlet Water Temperature (°C)	Range—Fan Frequency (Hz)	Range—Air Flow Velocity (m/s)
5–9	35–75	50–80	2–5
Interval—Water Mass Flow Rate (L/min)	Interval—Inlet Water Temperature (°C)	Interval—Fan Frequency (Hz)	
0.5	5	10	

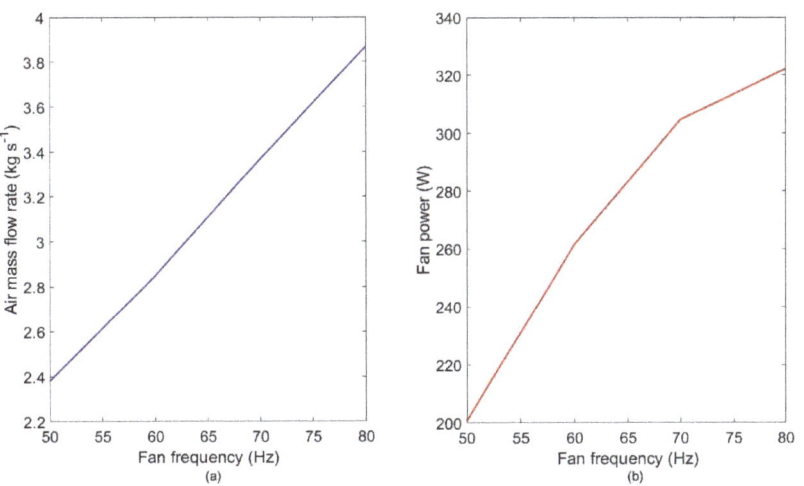

Figure 3. (a) Air mass flow rate depending on fan frequency; (b) fan power as a function of fan frequency.

Using Equations (20)–(24), the experimental overall heat transfer coefficient was calculated by:

$$Q = m_w \times \bar{c}_{p_w} \times \Delta T_w \tag{20}$$

$$T_{air,o} = \frac{Q}{m_{air} \times \bar{c}_{p_{air}}} + T_{air,i} \tag{21}$$

$$\Delta T_{LMTD} = \frac{\Delta T_2 - \Delta T_1}{\left(\frac{\Delta T_2}{\Delta T_1}\right)} \tag{22}$$

$$\Delta T_2 = T_{w,i} - T_{air,o} \text{ and } \Delta T_2 = T_{w,o} - T_{air,i} \tag{23}$$

$$U = \frac{Q}{A_R \times \Delta T_{LMTD}} \tag{24}$$

2.3. Model Application

The validated model was finally applied to define an efficient and reliable gas cooler design, as well as to evaluate its performance for different off-design conditions. Within scope, the on-design analysis investigated different designs of heat exchangers, such as the number of rows and length of the tube. The target of the off-design analysis was to evaluate the operation of the gas cooler in different ambient temperatures.

2.3.1. On-Design

The developed model was utilized to design an air-finned CO_2 gas cooler. Therefore, three and four numbers of rows and finned tubes with inner diameters of 6.85, 16, and 22.5 mm were considered. In order to identify an efficient gas cooler design, a compromise between the bundle area and the refrigerant pressure drop of the gas cooler was aimed for. The model was provided with the inlet, outlet temperatures and pressures, medium mass flows, and the duty of the heat exchanger as input variables. In addition, the design properties of the tube were specified, so the total area of the tube was calculated. The air and refrigerant heat transfer coefficients, which were initially unknown, were assumed, and then the overall heat transfer coefficient was calculated from Equation (1). The required exchanged area was calculated from Equation (2), and so the required number of tubes was obtained. In the iteration process, the air and refrigerant heat transfer coefficients were updated using Equations (3)–(17). The iteration was continued until the relative tolerance of the overall heat transfer coefficient for two continuous iterations was equal to 0.001.

The pressure drop was calculated by the following equation:

$$\Delta P = f \times \frac{G_{refri}^2}{2 \times \rho_{refri}} \times \frac{L}{d_i} \tag{25}$$

$$f = [1.82 \ln(Re_b - 1.64)]^{-2} \tag{26}$$

which Filonenko [27] applies for $10^4 \leq Re_b \leq 5 \times 10^6$.

In case of $Re_b \leq 10^4$, Blasius [27] correlations was applied:

$$f = \frac{0.316}{Re_b^{\frac{1}{4}}} \tag{27}$$

The thermophysical properties of air and refrigerant like density, viscosity, specific heat capacity, and thermal conductivity were obtained from REFPROP version 10 database. The model used the mean temperature and pressure of the mediums in order to calculate the heat transfer coefficients.

2.3.2. Off-Design

The calculations were used to evaluate the overall performance of the gas cooler in different conditions. The defined gas cooler was investigated in different ambient temperatures. The boundary conditions and operational parameters for the off-design analysis were defined according to Table 3.

Table 3. Off-design boundary conditions.

T_{amb} (°C)	$T_{refri,i}$ (°C)	$T_{refri,o}$ (°C)	P_{refri} (bar)	$\dot{m}_{refri}$ (kg/h)	$\dot{m}_a$ (kg/h)
20	78.3	20.02	93	184	93.15
21	79.3	21.03	93	184	92.79
22	80.3	22.03	93	184	92.29
23	81.2	23.04	93	184	91.80
24	82.2	24.05	93	184	91.30
25	83.2	25.07	93	184	90.60
26	84	26.12	93	208	101.67
27	84.7	27.2	93	233	112.59
28	88.6	28.31	93	257	125.66
29	85.9	29.48	93	282	132.37
30	86	30.71	93	306	141.14
31	86	32.01	93	330	148.68
32	85.8	33.52	93	379	165.14
33	85.5	35.09	93	428	179.17
34	85.3	36.61	93	476	190.51
35	85.1	38	93	525	199.50

Based on the off-design data, the overall heat transfer coefficient was calculated using Equations (3)–(17). In addition, investigation of different potential heat exchanger designs was conducted by comparing their overall heat transfer coefficient and the refrigerant pressure drop. The potential heat exchanger's designs were based on the on-design analysis, and were chosen in terms of pressure losses and bundle area. The selected air-cooled HEXs were designed with identical finned-tube properties.

3. Results and Discussion

3.1. Validation of the Model

An experimental campaign was carried out in order to validate the mathematical calculations for the performance of an air-finned CO_2 gas cooler. The U-value from experimental results was compared with the U-value obtained from the model. The following results were obtained using the proposed correlation from VDI-Heat Atlas [25] and Gnielinski's [26] correlation for the air-, and refrigerant-side, respectively. Figure 4a illustrates that most of the deviations were lower than 10%. Particularly, the average absolute deviation for 45 °C inlet water temperature and 50 Hz was 5%, while the maximum absolute deviation was 12%. The average absolute deviation for 50 Hz for different inlet water temperature was 11%. The calculations seem to approach the experimental results from 50 to 80 Hz. Figure 4b depicts that the absolute deviations for 80 Hz were even lower than 10%. Particularly, the absolute average deviations for 45 °C of 80 Hz was 5%, while the maximum absolute deviation was 9%. The absolute average deviation for 80 Hz of fan frequency for different water inlet temperatures was 6%.

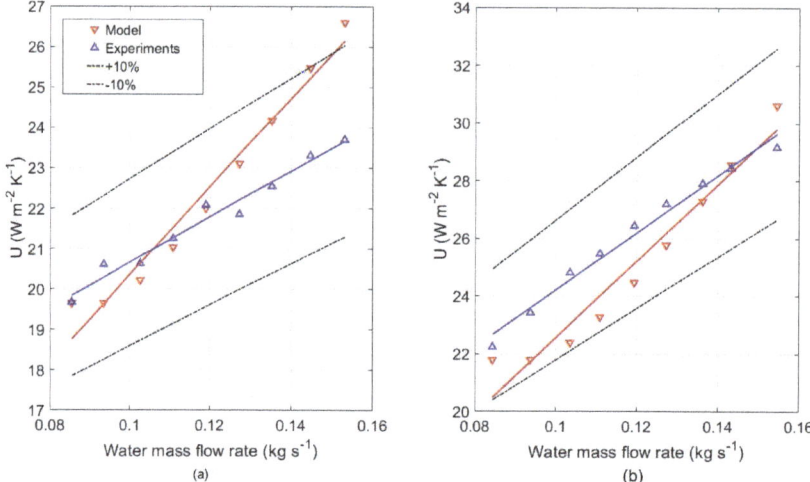

Figure 4. Deviations between the model and experimental overall heat transfer coefficient for 45 °C water inlet temperature at (**a**) 50 Hz; and (**b**) 80 Hz fan frequency.

3.2. Sensitivity Analysis

The most satisfying results were obtained by using the correlations proposed by VDI-Heat Atlas [25] and by Gnielinski [26] for the air- and the refrigerant-side heat transfer, accordingly. Fundamental investigations regarding the dependence of the heat transfer characteristics (U-value) on different operational parameters and applied correlations were conducted. Particularly, the impact of the fan frequency, the water mass flow, and the water inlet temperature on the performance of the air-cooled HEX were investigated. The increase of both the water mass flow and the fan frequency enhanced the overall performance of the heat exchanger, as Figure 5a illustrates. The increase of the water mass flow rate had a strong impact on the overall heat transfer coefficient by 20% and 30% of 50 and 80 Hz, accordingly. The increase of the fan frequency showed an average increase of the overall heat transfer coefficient of 8%, 6%, and 6% in the ranges of 50–60, 60–70 and 70–80 Hz, accordingly. The increase of the water inlet temperature affected in a similar way the overall heat transfer coefficient of the heat exchanger Figure 5b. An average increase of 35% of the overall heat transfer coefficient from 35 to 75 °C of water inlet temperature for fan frequency of 60 Hz was revealed.

The investigation of different heat transfer correlations is considered necessary to identify the most suitable air-side heat transfer correlation. The comparison between the refrigerant-side heat transfer correlations shows that the deviations are approximately 1% for all the different fan frequencies, while the correlation by Gnielinski [26] calculates the heat transfer as lower than by that of Dittus–Boelter in Figure 6a. The comparison between the air-side heat transfer correlations shows that the deviations are highly affected by the fan frequency. Particularly, absolute average deviations for 50 and 80 Hz are 8% and 4%, accordingly, while the maximum absolute deviations are 10% and 5%. It is revealed that with the increase of the fan frequency, the deviations between the correlations become lower, as shown in Figure 6b.

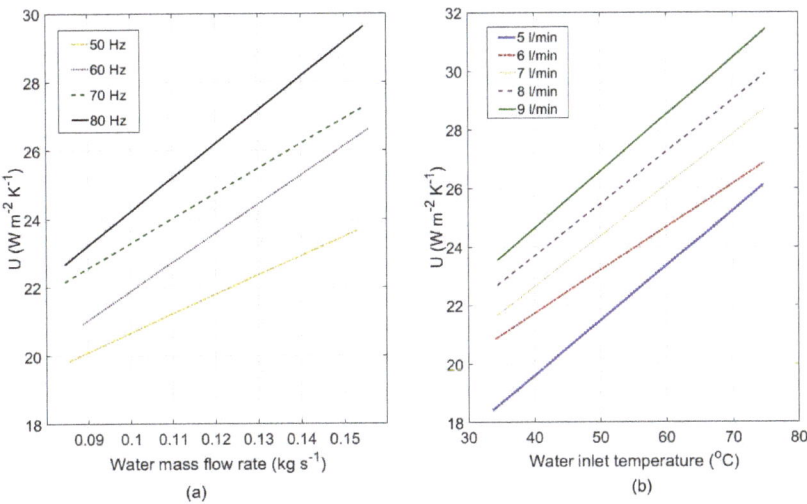

Figure 5. Overall heat transfer coefficient of (**a**) varying mass flow rate and fan frequencies; and (**b**) varying water inlet temperatures and water mass flow rates.

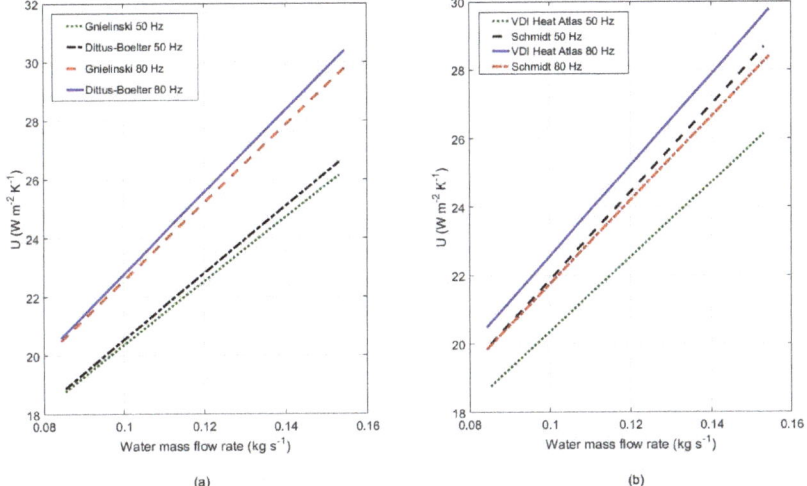

Figure 6. Comparison for 45 °C water inlet temperature for (**a**) refrigerant-side heat transfer correlations; and (**b**) air-side heat transfer correlations.

3.3. Design Procedure

The discussed model was applied. Thus, an on-design analysis was investigated in order to identify an efficient and reliable gas cooler in terms of pressure losses and required exchange area. The mathematical calculations were applied to on-design analysis using the correlations by Gnielinski [26] and those proposed by VDI-Heat Atlas [25] for the refrigerant- and air-side heat transfer characteristics, accordingly. The investigation of different number of rows (NR) and finned-tubes showed that the heat exchanger with four rows and the smallest diameter has a smaller bundle area (Figure 7).

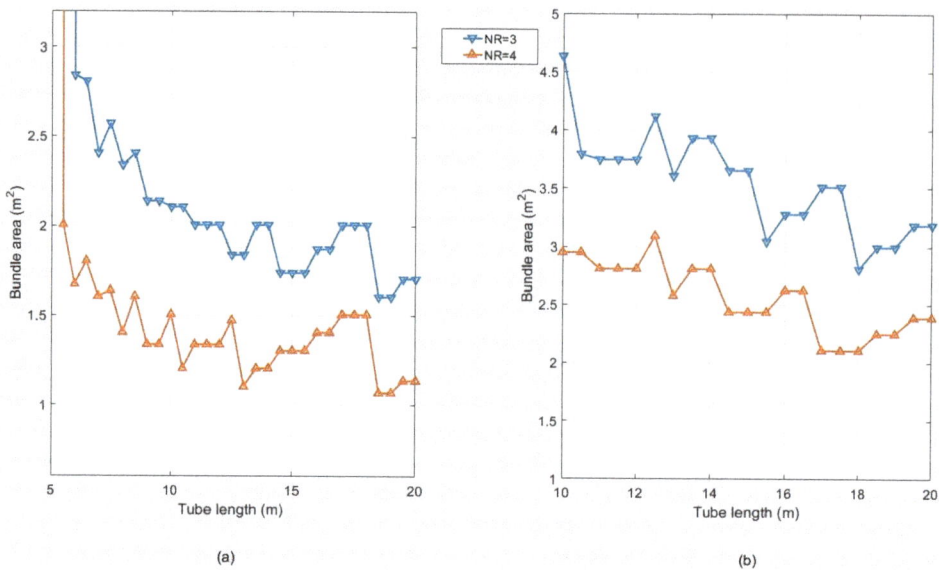

Figure 7. Bundle area vs. tube length for 3 and 4 number of rows (**a**) outside diameter of 7.35 mm; and (**b**) outside diameter of 18.5 mm.

Taking the pressure drop into consideration, the results show that the tube with the outer diameter of 7.35 mm cannot be used, as the occurring pressure drop causes serious operation problems to the system, due to pressure drops exceeding 0.3 bar (Figure 8a). Instead, the heat exchanger designed with the tube with outside diameter of 18.5 mm can be used for lengths to 20 m (see Figure 8b). Figures 9 and 10 show the Reynolds number for air- and refrigerant-side, respectively.

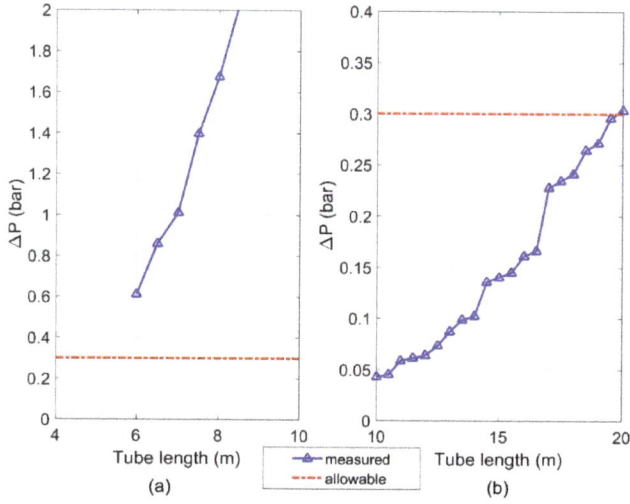

Figure 8. Refrigerant pressure drop vs. length of the tube for 4 rows (**a**) outside diameter of 7.35 mm; and (**b**) outside diameter of 18.5 mm.

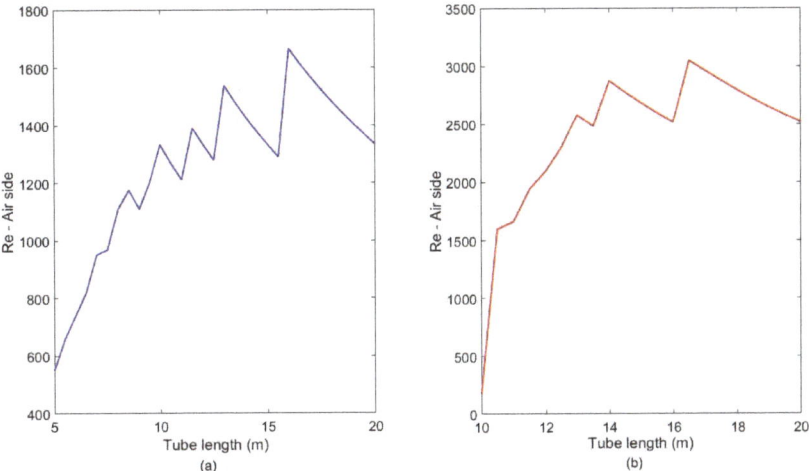

Figure 9. Reynolds number air side (**a**) outside diameter of 7.35 mm; and (**b**) outside diameter of 18.5 mm.

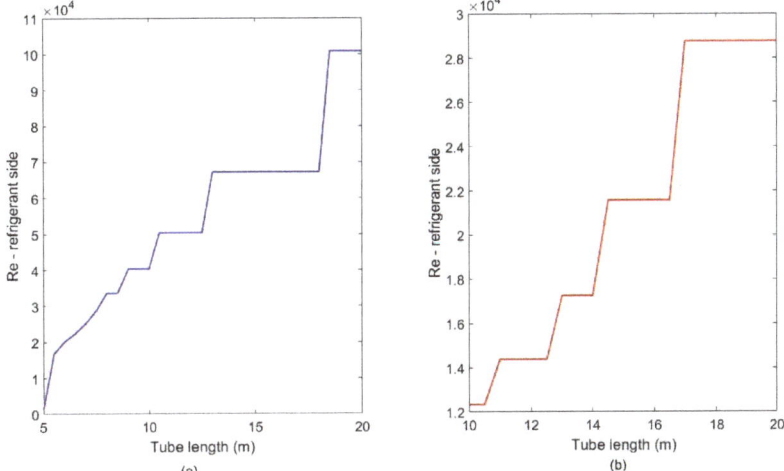

Figure 10. Reynolds number refrigerant side (**a**) outside diameter of 7.35 mm; and (**b**) outside diameter of 18.5 mm.

Based on the presented results, the final design of the air-finned CO_2 gas cooler is defined for a finned-tube (type-*U*), according to the specifications listed in Table 4.

3.4. Evaluation at Off-Design Conditions

The off-design analysis of the defined gas cooler was carried out for different ambient temperatures from 20 to 35 °C, and a comparison between potential gas coolers was made. The potential heat exchangers were chosen in terms of pressure losses and bundle area, while they had identical finned tube properties, except for the length of the tube. Particularly, the gas cooler with bundle area of 2.39 m^2 was characterized by its high pressure losses of 0.3 bar, while the gas cooler with bundle area of 2.81 m^2 showed relatively low pressure losses of 0.06 bar. The gas cooler with bundle area of 2.43 m^2

was chosen, due to its combination of comparatively low pressure losses of 0.14 bar, as well as its low bundle area.

Table 4. Gas cooler technical specifications.

Gas Cooler Specifications	
Total length of the tube (m)	17
Number of passes per tube (-)	14
Number of passes (-)	43
Length of each pass (m)	1.2
Number of rows (-)	4
Number of tubes per row (-)	3
Bundle area (m^2)	2.11
Refrigerant pressure drop (bar)	0.23
Air Finned Tube Specifications	
Outside diameter (mm)	18.5
Wall thickness (mm)	1.25
Fin height (mm)	10

Figure 11a below shows that the defined gas cooler operates better than the other air-cooled HEX in different ambient temperatures. The defined gas cooler's overall heat transfer coefficient was remarkably higher than the heat exchangers with the bundle area of 2.43 and 2.81 m^2. On the other hand, the U-value of the heat exchanger of 2.39 m^2 bundle area is near to the defined heat exchanger, (A_b = 2.11 m^2) but the refrigerant pressure drop is significantly higher (Figure 11b).

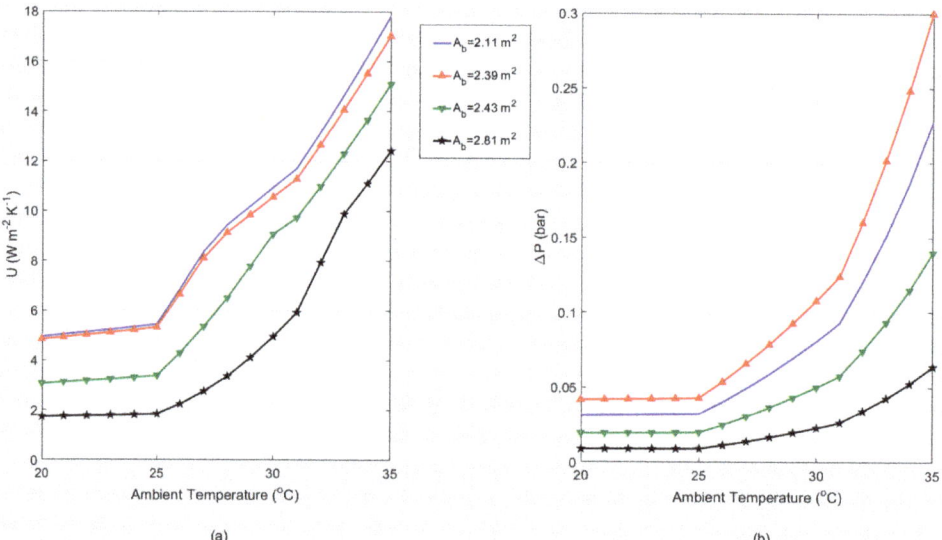

Figure 11. Comparison between potential heat exchangers in different ambient temperatures of (**a**) the overall heat transfer coefficient; and (**b**) the refrigerant pressure drop.

4. Conclusions

A design procedure for a CO_2 finned-tube gas cooler was developed and validated experimentally. The experimental focus was laid on the validation of the air-side heat transfer. Absolute deviations between the model and the experiments were extracted and prove the reliability of the selected heat transfer correlations. The developed simulation model was used to design an efficient gas cooler and evaluate its performance in different ambient temperatures. Based on these results, the following conclusions are justified:

The deviations between the calculations and the experiments are highly affected by the fan frequency, the water mass flow, and the water inlet temperature. It was found that the absolute average deviations for 60–80 Hz are less than 10%. The increase in the fan frequency, the water mass flow, and the water inlet temperature caused an improvement of the overall heat transfer coefficient of the heat exchanger. The comparison between potential heat transfer correlations showed that the combination of the correlations proposed by VDI-Heat Atlas [25] and by Gnielinski [26] for the air- and refrigerant-side heat transfer coefficient, respectively, approached the experimental results better.

The heat exchanger with four rows has a smaller bundle area than with three rows, while the investigation of different tubes showed that the heat exchanger designed with the smallest diameter has the smallest bundle area and the highest refrigerant pressure drop. As a compromise between the bundle area and the refrigerant pressure drop, a gas cooler of 2.11 m^2 and refrigerant pressure drop of 0.23 bar was defined. Comparison between potential gas coolers was made, and the results show that the defined gas cooler operates more efficiently for different ambient conditions, compared to other potential heat exchangers.

Due to the pressure limitations of the equipment, the experimental validation of the model was conducted with water as working fluid. This procedure enables a reliable validation of the applied heat transfer correlation at the air side. However, the use of CO_2 as working fluid in the tubes would improve validation approach. Therefore, experiments with CO_2 are suggested for further work, next to the off-design analysis for the entire air conditioning system including the developed gas cooler model.

Author Contributions: All authors contributed to this work by collaboration. Methodology, software, validation, investigation and writing—original draft preparation: C.A.; off-design evaluation: O.A.; conceptual design of the study and review and editing: F.H.; methodology, software and writing, and review and editing: T.C.R.; supervision: S.K. and D.B. All authors have read and agreed to the published version of the manuscript.

Funding: This publication is strongly related to a research project (Project-ID TEW01CO2P-73707) at the Center of Energy Technology (ZET) of the University of Bayreuth, which is financed by the Bavarian State Ministry of Environment and Consumer Protection. The main research work of this publication was conducted during a stay of Charalampos Alexopoulos in Bayreuth, which was funded by the ERASMUS+ Programme. Additionally, the fourth author would like to state that this scientific paper was supported by the Onassis Foundation; Scholarship ID: G ZO 025-1/2018-2019.

Conflicts of Interest: The authors declare no conflicts of interest.

Nomenclature

A	Area	m^2
$\bar{c}_p$	Mean specific heat capacity	J kg^{-1} K^{-1}
d	Diameter	m
f	Friction factor	-
G	Mass velocity	kg m^2 s^{-1}
h	Heat transfer coefficient	W m^{-2} K^{-1}
H	Height	m
k	Thermal conductivity	W m^{-1} K^{-1}
L	Length	m
m	Mass flow rate	kg s^{-1}
Nt	Number of tubes	-
Nu	Nusselt number	-
P	Pressure	bar
Pr	Prandtl number	-
Q	Heat transfer rate	W
Re	Reynolds number	-

s	Spacing	m
t	Thickness	m
T	Temperature	K
U	Overall heat transfer coefficient	$W\,m^{-2}\,K^{-1}$
w	Velocity	$m\,s^{-1}$

Greek symbols

η	Efficiency	-
μ	Dynamic viscosity	$kg\,m^{-1}\,s^{-1}$
ρ	Density	$kg\,m^{-3}$

Subscripts

air	Air
b	Bundle
f	Fin
i	Inner
inf	Inflow
o	Outer
R	Required
refri	Refrigerant
s	Smallest cross-sectional
to	Bare tube without fins
w	Water

Abbreviations

CFCs	Chlorofluorocarbons
CO_2	Carbon dioxide
COP	Coefficient of performance
GWP	Global warming potential
HCFCs	Hydrochlorofluorocarbons
HEX	Heat exchanger
HFCs	Hydrofluorocarbons
HVAC	Heating, ventilation and air conditioning
LMTD	Logarithmic mean temperature difference
NR	Number of rows
ODP	Ozone depletion potential

References

1. Hawken, P. (Ed.) *Drawdown: The Most Comprehensive Plan Ever Proposed to Reverse Global Warming*, 1st ed.; Penguin Books: New York, NY, USA, 2017; ISBN 978-0-14-313044-4.
2. Schulz, M.; Kourkoulas, D. Council of the European Union Regulation (EU) No 517/2014 of the European Parliament and of the Council of 16 April 2014 on fluorinated greenhouse gases and repealing Regulation (EC) No 842/2006. *Off. J. Eur. Union* **2014**, *2014*, L150.
3. *General Office of the State Council. China State Council 2014–2015 Energy Conservation, Emissions Reduction and Low Carbon Development Action Plan*; The General Office of the State Council: Beijing, China, 2014.
4. *California Air Resources Board Draft Short-Lived Climate Pollutant Reduction Strategy*; California Environmental Protection Agency: Sacramento, CA, USA, 2015.
5. David, A.; Mathiesen, B.V.; Averfalk, H.; Werner, S.; Lund, H. Heat Roadmap Europe: Large-Scale Electric Heat Pumps in District Heating Systems. *Energies* **2017**, *10*, 578. [CrossRef]
6. Braimakis, K.; Roumpedakis, T.C.; Leontaritis, A.-D.; Karellas, S.; Roumpedakis, T.C.; Leontaritis, A.-D.; Karellas, S. Comparison of Environmentally Friendly Working Fluids for Organic Rankine Cycles. In *Advances in New Heat Transfer Fluids*; Minea, A.A., Ed.; CRC Press: Boca Raton, FL, USA, 2017; pp. 377–426.

7. Gambhir, A.; Napp, T.; Hawkes, A.; Höglund-Isaksson, L.; Winiwarter, W.; Purohit, P.; Wagner, F.; Bernie, D.; Lowe, J. The Contribution of Non-CO2 Greenhouse Gas Mitigation to Achieving Long-Term Temperature Goals. *Energies* **2017**, *10*, 602. [CrossRef]
8. Tao, Y.B.; He, Y.L.; Tao, W.Q. Exergetic analysis of transcritical CO2 residential air-conditioning system based on experimental data. *Appl. Energy* **2010**, *87*, 3065–3072. [CrossRef]
9. Shah, R.K.; Sekulic, D.P. *Fundamentals of Heat Exchanger Design*; John Wiley & Sons Inc.: Hoboken, NJ, USA, 2003; ISBN 978-0-471-32171-2.
10. Lorentzen, G.; Pettersen, J. A new, efficient and environmentally benign system for car air-conditioning. *Int. J. Refrig.* **1993**, *16*, 4–12. [CrossRef]
11. Pitla, S.S.; Groll, E.A.; Ramadhyani, S. Convective Heat Transfer from In-Tube Cooling of Turbulent Supercritical Carbon Dioxide: Part 2—Experimental Data and Numerical Predictions. *HVAC&R Res.* **2011**, *7*, 367–382. [CrossRef]
12. Son, C.-H.; Park, S.-J. An experimental study on heat transfer and pressure drop characteristics of carbon dioxide during gas cooling process in a horizontal tube. *Int. J. Refrig.* **2006**, *29*, 539–546. [CrossRef]
13. Zhang, H.; Cheng, K.; Huai, X.; Guo, J. Performance analysis of printed circuit heat exchanger for supercritical carbon dioxide and water. *IOP Conf. Ser. Mater. Sci. Eng.* **2020**, *721*, 012039. [CrossRef]
14. Jadhav, N.P.; Deshmukh, S.; Lele, M.M. Numerical Simulation of Fin and Tube Gas Cooler for Transcritical CO2 Air Conditioning System. *Int. J. Eng. Res. Technol.* **2012**, *1*. Available online: https://www.ijert.org/research/numerical-simulation-of-fin-and-tube-gas-cooler-for-transcritical-co2-air-conditioning-system-IJERTV1IS10421.pdf (accessed on 15 April 2020).
15. Liu, Z.-B.; He, Y.-L.; Yang, Y.-F.; Fei, J.-Y. Experimental study on heat transfer and pressure drop of supercritical CO_2 cooled in a large tube. *Appl. Therm. Eng.* **2014**, *70*, 307–315. [CrossRef]
16. Jiang, P.-X.; Zhao, C.-R.; Shi, R.-F.; Chen, Y.; Ambrosini, W. Experimental and numerical study of convection heat transfer of CO2 at super-critical pressures during cooling in small vertical tube. *Int. J. Heat Mass Transf.* **2009**, *52*, 4748–4756. [CrossRef]
17. Chai, L.; Tsamos, K.M.; Tassou, S.A. Modelling and Evaluation of the Thermohydraulic Performance of Finned-Tube Supercritical Carbon Dioxide Gas Coolers. *Energies* **2020**, *13*, 1031. [CrossRef]
18. Cheng, L.; Ribatski, G.; Thome, J.R. Analysis of supercritical CO_2 cooling in macro- and micro-channels. *Int. J. Refrig.* **2008**, *31*, 1301–1316. [CrossRef]
19. Ge, Y.T.; Cropper, R.T. Simulation and performance evaluation of finned-tube CO_2 gas coolers for refrigeration systems. *Appl. Therm. Eng.* **2009**, *29*, 957–965. [CrossRef]
20. Marcinichen, J.B.; Thome, J.R.; Pereira, R.H. Working fluid charge reduction. Part II: Supercritical CO2 gas cooler designed for light commercial appliances. *Int. J. Refrig.* **2016**, *65*, 273–286. [CrossRef]
21. Zilio, C.; Cecchinato, L.; Corradi, M.; Schiochet, G. An Assessment of Heat Transfer through Fins in a Fin-and-Tube Gas Cooler for Transcritical Carbon Dioxide Cycles. *HVAC&R Res.* **2007**, *13*, 457–469. [CrossRef]
22. Gupta, D.K.; Dasgupta, M.S. Simulation and performance optimization of finned tube gas cooler for trans-critical CO_2 refrigeration system in Indian context. *Int. J. Refrig.* **2014**, *38*, 153–167. [CrossRef]
23. Santosa, I.M.C.; Tsamos, K.M.; Gowreesunker, B.L.; Tassou, S.A. Experimental and CFD investigation of overall heat transfer coefficient of finned tube CO_2 gas coolers. *Energy Procedia* **2019**, *161*, 300–308. [CrossRef]
24. *MATLAB and Statistics Toolbox Release*; The MathWorks, Inc.: Natick, MA, USA, 2019.
25. Schmidt, K.G. M1 Heat Transfer to Finned Tubes. In *VDI Heat Atlas*; *VDI-Buch*; Springer: Berlin/Heidelberg, Germany, 2010; pp. 1273–1278. ISBN 978-3-540-77876-9.
26. Gnielinski, V. Neue Gleichungen für den Wärme- und Stoffübergang in turbulent durchströmten Rohren und Kanälen. *Forsch. Ing.* **1975**, *41*, 8–16. (In Germany) [CrossRef]
27. Dittus, F.W.; Boelter, L.M.K. Heat transfer in automobile radiators of the tabular type. *Univ. Calif. Publ. Eng.* **1930**, *2*, 443–461.
28. Schmidt, E.F. Wärmeübergang und Druckverlust in Rohrschlangen. *Chem. Ing. Tech.* **1967**, *39*, 781–789. (In Germany) [CrossRef]

© 2020 by the authors. Licensee MDPI, Basel, Switzerland. This article is an open access article distributed under the terms and conditions of the Creative Commons Attribution (CC BY) license (http://creativecommons.org/licenses/by/4.0/).

Article

Feasibility Study of a Centralised Electrically Driven Air Source Heat Pump Water Heater to Face Energy Poverty in Block Dwellings in Madrid (Spain)

Roberto Barrella, Irene Priego, José Ignacio Linares *, Eva Arenas, José Carlos Romero and Efraim Centeno

ICAI School of Engineering, Comillas Pontifical University, 28015 Madrid, Spain;
roberto.barrella@iit.comillas.edu (R.B.); 201807511@alu.comillas.edu (I.P.); earenas@icai.comillas.edu (E.A.); jose.romero@iit.comillas.edu (J.C.R.); efraim.centeno@iit.comillas.edu (E.C.)
* Correspondence: linares@comillas.edu

Received: 16 April 2020; Accepted: 25 May 2020; Published: 28 May 2020

Abstract: Energy poverty can be defined as the inability to pay the bills that are required for maintaining the comfort conditions (usually in winter) in dwellings. The use of energy efficient systems is one way forward to mitigate this problem, with one option being the electrically driven air source heat pump water heater. This paper assesses the performance of a centralised heat pump (200 kW of heating capacity) to meet the space heating demand of block dwellings in Madrid (tier four out of five in winter severity in Spain). Two models have been developed to obtain the following variables: the hourly thermal energy demand and the off-design heat pump performance. The proposed heat pump is driven by a motor with variable rotational speed to modulate the heating capacity in an efficient way. A back-up system is also considered to meet the peak demand. A levelised cost of heating of 92.22 €/MWh is obtained for a middle-level energy efficiency in housing (class E, close to D). Moreover, the following energy-environmental parameters have been achieved: more than 74% share of renewable energy in primary energy and 131.7 g CO_2 avoided per kWh met. A reduction of 60% in the heating cost per dwelling is obtained if an energy retrofit is carried out, improving the energy performance class from E to C. These results prove that the proposed technology is among the most promising measures for addressing energy poverty in vulnerable households.

Keywords: energy poverty; centralised heat pump; hourly heating demand; off-design heat pump model

1. Introduction

Although there is no agreed definition of energy poverty, there is some consensus in identifying it as the situation suffered by "individuals or households that are unable to adequately heat, cool or provide other necessary energy services in their homes at an affordable cost" [1]. Approximately 40 million people in the Union (8% of the total population) suffer from this situation, according to data from the latest European Living Conditions Survey [2,3]. In Spain, the incidence ranges from 7.2% to 16.9% of the population [4], depending on the indicator used, according to the latest update of the National Strategy against Energy Poverty.

In this context, there is a wide consensus in recognising that energy poverty is a key social challenge that must be addressed by Member States through their public policies. The private sector can also contribute to tackle this issue, by means of social entrepreneurship and the involvement of large corporations through their Corporate Social Responsibility [5].

Energy poverty, indeed a manifestation of general poverty, is directly linked to low income, and many low-income households are energy poor. Nevertheless, energy poverty does not fully overlap with income poverty. Two energy-related variables, namely housing energy efficiency and energy prices, are also key features characterising the problem [6].

In terms of consequences, focusing only on health impacts, it is well documented that living in households with inadequate heating or cooling has detrimental consequences for respiratory, circulatory and cardiovascular systems, as well as for mental health and well-being [7]. Additionally, energy poverty impacts on other aspects of people life, namely economic, social, and educational ones, among others [8]. In the thick of energy needs that remain totally or partially unmet in vulnerable households, heating needs are particularly noteworthy. A potential improvement in this respect could be the installation of clean and efficient heating systems, such as the centralised heat pump analysed in this paper.

The air source heat pumps are a promising alternative to replace the old heating installations of block dwellings suffering energy poverty. These dwellings usually have oversized radiators, which enables them to be fed with low temperature water, then having a temperature range that is suitable for air source heat pumps water heaters (ASHPWH) [9]. European Union considers that the energy taken from the air in the evaporator can be accounted as renewable energy if the heating seasonal performance factor exceeds certain values [10]. Due to this fact, the replacement of centralised gas boilers by heat pumps makes it possible to introduce renewable energies in heating, so achieving an efficient heating system more de-carbonised than the former boiler. Such replacement should take the impact of the electricity taken from the grid in terms of CO_2 emissions, which depends on the electricity generation mix of the country, into account. Furthermore, heat pumps must face two specific environmental issues, namely, the ozone deployment potential (solved some time ago with the hydrofluorocarbons, HFC) and the global warming potential (GWP). In fact, the GWP of the HFCs used as refrigerants is usually too high. European Union has regulated the use of fluorinated greenhouse gases (F-gases) [11], establishing an agenda to move from HFC to natural refrigerants (propane, butane, ammonia, CO_2, etc.), passing through hydrofluoroolefins (HFO) as intermediate fluids. Spanish regulation [12] recently included heat pumps as a renewable option to supply thermal services to buildings, especially domestic hot water. Despite this, the use of heat pumps for heating purposes is far away from common practice in Spain. Conversely, space heating from heat pumps is usually a by-product of cooling services, using the so-called reversible heat pumps. This situation is even considered by the European Union [10], which penalises reversible heat pumps when counting renewable energy due to the fact that they are usually designed to operate in cooling mode.

Curve fitting from actual machines is a common methodology for modeling the behaviour of heat pumps. Accordingly, Underwood et al. [13] develop a model that is based on refrigerant-side variables, which makes it suitable for the analysis of the performance of heat pumps in service. So, this type of model is used when the focus is on the representation of the entire system building-heat pump. For instance, Lohani et al. [14] integrate correlation curves of coefficient of performance (COP) for ground and air source heat pumps in a building modelling software that lacks heat pump models. In some cases [15], the curve fitting is based on the theoretical behaviour of Carnot efficiency, as compared with actual performance. These models are used to identify key performance parameters in a site, as carried out by Vieira et al. [16].

Physical models of heat pumps are based on energy and heat transfer equations of the different components. The key components are the heat exchangers, which are modelled using the effectiveness-number of transfer units method or the equivalent logarithmic mean temperature difference method [17]. These types of methods usually consider the single phase and phase change zones. For example, Fardoun et al. [18] propose a quasi-steady state model that is based on an iterative procedure for the heat exchangers. In this sense, Patnode [19] develops a model of heat exchangers that is based on the Dittus–Boelter correlation, which can obtain the overall heat transfer coefficient as a function of the mass flow rate. This type of models is the base of rules and standards that determine the COP of ASHPWH as a function of water temperature and air dry and wet temperature [20].

Other studies use dynamic models, which are usually based on commercial software as TRNSYS. This type of analyses might pursue the improvement of the control strategies in the operation of heat pumps [21] or enhance the coupling with the thermal inertia of the building [22]. In short, dynamic models are useful when the seasonal performance is sought [23].

The thermal load, which is required as an input data to simulate the behavior of the heat pump, can be calculated using hourly average data or load predictions. Xu et al. [24] carry out a case study while using data that were recorded along two years of operation of a data center, evaluating the performance of the combined cooling, heat, and power (CCHP) system, also performing transient modeling. Gadd et al. performed an analysis of the heat meter readings that were obtained on an hourly basis along one year [25], aiming to provide heat load patterns, whereas Bacher et al. [26] attempted to develop a method for predicting the space heating thermal load in a dwelling. The latter model is based on measured data from actual houses in combination with local climate measurements and weather forecasts. Noussan et al. [27] propose a thermal demand model built up while using heat measurements taken every six minutes along several years of operation. Energy Plus software from the U.S. Department of Energy [28] was used by many researchers to determine the thermal load. In this sense, Wood et al. [29] and Michopoulos et al. [30] employed this software to analyse the use of biomass in space heating. Other authors used regressive models to obtain thermal energy demand prediction methods [31] or autoregression analysis with exogenous time and temperature indexes [32]. In this sense, Powell et al. [33] selected nonlinear autoregressive models with exogenous inputs as the best methodology based on artificial neural networks.

Several scholars used the degree-day method to obtain the hourly thermal energy demand profile. For example, Büyükalaca et al. [34] estimate the energy needs in a building in Turkey by calculating the heating and cooling degree-days using variable-base temperatures. Furthermore, Martinaitis et al. [35] perform an exergy analysis of buildings based on the degree-days method. Moreover, Layberry [36] analysed the errors in the degree-day method that may affect the building energy demand analysis. Carlos et al. [37] combined solar radiation data with the degree-day method and compared several different simplified methodologies for building energy performance assessment in winter. In Spain, the winter climatic severity index is defined from the winter degree-days and solar radiation measurements [38]. This index is used to determine the thermal energy demand and it makes it possible to characterise the country's climatic zones in order to assess the energy requirements, according to the building energy performance regulations [39]. This thermal energy demand modelling is also suitable for assessing the performance of other thermal devices for the evaluated scenario.

This paper analyses the efficiency of an electrically driven a heat pump as a realistic alternative to achieve winter thermal comfort in vulnerable households' dwellings. Thermally driven heat pumps (driven by absorption or internal combustion engines) also constitute a feasible option; nevertheless, this research is focused on electrically driven heat pumps, due to its higher commercial deployment in Spain. In order to do so, the baseline case is defined for a block of dwellings with a middle-level of energy efficiency and, additionally, retrofitting alternatives are considered for under average situations. The performance of the heat pump in terms of cost and CO_2 emissions is compared with other alternatives (centralized and decentralized boilers). The cost breakdown of the heat pump is detailed, allowing for the evaluation of subsidy schemes in order to fund this kind of devices.

The novelty of this paper lies on the assessment of the use of centralised electrically driven heat pumps to meet the heating demand in Madrid. Such solutions are usually employed in northern Europe areas, especially with ground source heat pumps. However, the use of air source heat pumps is not common in Spain, as it can be followed from the support to these devices in the last release of the Spanish Technical Building Code [12]. In this sense, the developed heat pump model is not sophisticated, although it is accurate enough, as can be derived from the comparison with a commercial machine. Regarding the heating demand forecasting model, a simpler version has been proposed by the authors [40], but, as a novelty, the version that is used in this manuscript is able to retain information regarding thermal insulation of the building. This information made it possible to assess the effect of energy building retrofitting. Therefore, the aim of this paper is to analyse the feasibility of ASHPWH as active measure to fight energy poverty in dwelling blocks. The environmental and economic assessment performed in this paper can eventually provide insightful information to policy makers for implementing clean and efficient measures to tackle energy poverty.

2. Methodology

2.1. Hourly Heating Demand

Building Technical Code (BTC) in Spain establishes a procedure for assessing the energy demand in both winter and summer as a function of the climate severity index (CSI) [41]. In this work, only heating demand is assessed, so focusing on winter energy needs, because energy poverty studies have been mainly focused on this season [42].

The reference specific demand (RD, kWh/m^2) in winter is given by Equation (1), where WSI stands for the winter severity index and Table 1 provides the coefficients α and β. The winter severity index is defined in Equation (2), where RAD is the average accumulated global radiation over horizontal surface during January, February, and December (Equation (3)), and DD is the average degree-days (at base temperature T_b = 20 °C) for the same months (Equation (4a)) [43]. Table 2 provides the coefficients for Equation (2). The calculation of RAD requires the global hourly radiation over horizontal surface (r_k), whereas the calculation of DD requires the hourly temperature difference (ΔT_k), as defined by Equation (4b). Hourly values of T_k are available for each climatic zone in the web site of the BTC [44].

$$RD = \alpha + \beta \cdot WSI \tag{1}$$

$$WSI = a \cdot RAD + b \cdot DD + c \cdot RAD \cdot DD + d \cdot RAD^2 + e \cdot DD^2 + f \tag{2}$$

$$RAD = \frac{\sum_{k=1}^{24 \times 90} r_k}{3} \tag{3}$$

$$DD = \frac{\sum_{k=1}^{24 \times 90} \Delta T_k}{24 \times 3} \tag{4a}$$

$$\Delta T_k = \begin{cases} T_b - T_k & if\ T_b > T_k \\ 0 & otherwise \end{cases} \tag{4b}$$

Table 1. Coefficients required to obtain the reference specific demand in winter [41].

	α	β
Single-family house	9.29	54.98
Block dwellings	3.51	39.57

Table 2. Coefficients required to obtain the winter severity index (WSI) [43].

a	b	c	d	e	f
-8.35×10^{-3}	3.72×10^{-3}	-8.62×10^{-6}	4.88×10^{-5}	7.15×10^{-7}	-6.81×10^{-2}

RD and WSI are overall values, that is, they are calculated for the complete winter season, as it is observed in Equations (1) and (2). Based on these equations, a Taylor series expansion of first order around (RAD, DD) has been carried out over WSI. This procedure leads to an hourly expression of the specific reference demand (Equation (5a)), where: (1) the index "j" is extended from 1 to 4368 (number of hours from January to March and from October to December), (2) N_d is equal to 182 days (number of days in the same period), (3) N_m is equal to 6 (number of months), and (4) the star denotes that this specific reference demand needs to be corrected. This correction has to be done to take into account two different issues. Firstly, the effect of the radiation and second, the fact that three additional months have been included with respect to the original correlation.

The correction of the radiation is performed because, sometimes, its value is high enough to result in a cooling demand (negative heating demand). In this case, the heating demand is set to zero. On the other hand, to consider the inclusion of additional months in the formulation, a reduction coefficient (C_r) is defined as the ratio of the actual specific seasonal demand (RD^a, given in regulations [45] and equal to 53 kWh/m² for Madrid) to the summation of RD_j^* over the 4368 h. Accordingly, the corrected hourly specific reference demand $\left(RD_j\right)$ is given in Equation (5d).

$$RD_j^* = \frac{\alpha + \beta \cdot (WSI - \rho \cdot RAD - \delta \cdot DD)}{24 \cdot N_d} + \left(\frac{\beta \cdot \rho}{N_m}\right) \cdot r_j + \left(\frac{\beta \cdot \delta}{24 \cdot N_m}\right) \cdot \Delta T_j \tag{5a}$$

$$\rho = a + 2 \cdot d \cdot RAD + c \cdot DD \tag{5b}$$

$$\delta = b + c \cdot RAD + 2 \cdot e \cdot DD \tag{5c}$$

$$RD_j = RD_j^* \cdot \underbrace{\left(\frac{RD^a}{\sum_{j=1}^{4368} RD_j^*}\right)}_{C_r} \cdot \begin{cases} 1 & \text{if } RD_j^* > 0 \\ 0 & \text{otherwise} \end{cases} \tag{5d}$$

Once the hourly specific reference demand is obtained, it should be corrected according to the energy performance index (*EPI*) and the ratio of the reference demand of the whole stock of reference buildings to the 10-th percentile of this stock (*R*) [41]. This correction leads to the calculation of the hourly absolute demand (D_j, Equation (6)), where the heated area (*A*) has been included. In Equation (6), the *EPI* is obtained from the energy performance certificate of the building and *R* is given at Table 3, where the climatic zone ranges from mild winter (A) to severe winter (E). For the current research, the *EPI* values have been calculated by cross-correlating the CENSUS 2011 data [46] and the buildings-energy-certification data [47]. The average *EPI* values for D zone (Madrid) are found to be 3.53 for block dwellings built before 1981, 2.18 if they were built between 1981 and 2007 and 0.92 for the ones built after 2008. Vulnerable households typically live in old buildings and use inefficient heating installations, typically electric radiators, as shown in the literature [48–50]. For this reason, the first two building age categories are those in which this collective of households is commonly located. In this study, as explained in Section 3.2, the block dwellings built between 1981 and 2007 are chosen as the baseline case.

$$D_j = A \cdot RD_j \cdot \left(\frac{1 + (EPI - 0.6) \cdot 2 \cdot (R - 1)}{R}\right) \tag{6}$$

Table 3. Values for R in Equation (6) [41].

Winter Climatic Zone	Single-Family House	Block Dwellings
A	1.7	1.7
B	1.6	1.7
C	1.5	1.7
D	1.5	1.7
E	1.4	1.7

Figure 1 displays the hourly demand for each ambient wet bulb temperature (Wet bulb temperature is used as a measurement of the enthalpy of humid air). It shows a cloud of points following a linear regression curve (enclosed in a red dashed line), along with a set of disperse data with lower heating demand. The data density increases with the temperature, in agreement with the radiation issue previously explained.

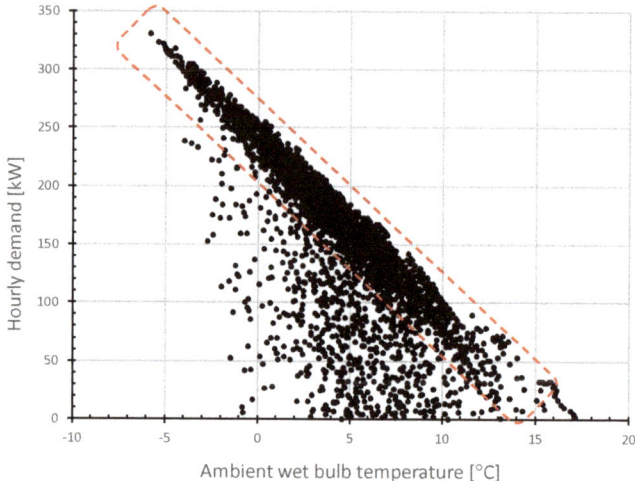

Figure 1. Hourly heating demand profile for each ambient wet bulb temperature for a set of block dwellings in Madrid built between 1981 to 2007 with 6000 m^2 of total heated surface.

Finally, the hourly demand is sorted from maximum to minimum, obtaining the annual cumulative heating demand profile, as shown in Figure 2.

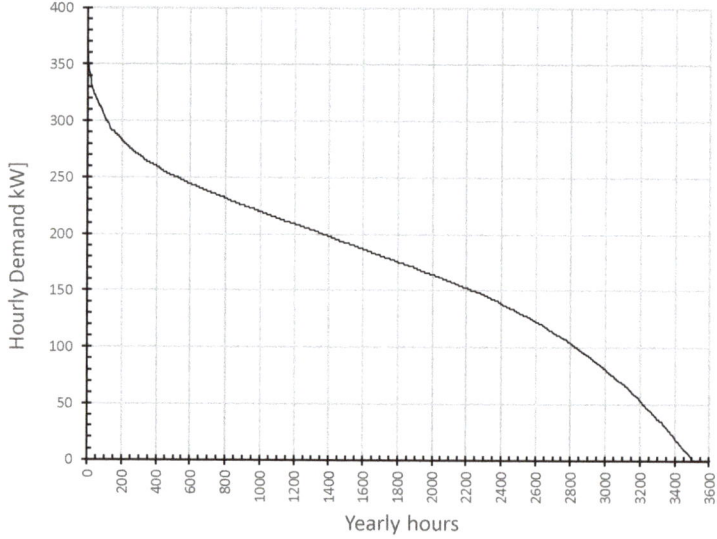

Figure 2. Annual cumulative heating demand profile for a set of block dwellings in Madrid built between 1981 to 2007 with 6000 m^2 of total heated surface.

2.2. Heat Pump Model

Two models have been developed for obtaining the performance of the heat pump: one for the best efficiency point (BEP) and another for the off-design operation. The former is used to size the main components and the latter to obtain the performance map. The heat pump uses the air as thermal source. Thus, a rotational speed control (inverter) driving the compressor motor is assumed to avoid the loss of heating capacity when the ambient temperature decreases. The nominal rotational speed is taken as 1490 rpm, with a range of variation from 745 to 2235 rpm (±50%). Out of these limits, a back-up system is necessary, assumed as a condensing boiler with modulation, fuelled by natural gas. The efficiency of this boiler has been taken as 95% (based on higher heating value, HHV), assumed constant when considering the usual seasonal performance factor values that were reported by manufacturers [51]. The rotational speed of the evaporator fan is also controlled to keep constant the temperature drop in the air to further improve the heat pump efficiency. The heat pump is an air/water system, so water of the existing radiators heating loop is heated in the condenser. The heat pump takes advantage of the usual oversizing employed in the radiators heat transfer area to make them behave as low temperature radiators, thus enabling the use of heat pump as heater. Domestic hot water demand is covered by solar thermal energy that is supported by natural gas, with this aspect being out of the scope of this analysis. Cooling demand is not considered, according to the common trend in energy poverty studies [42].

Figure 3 shows a scheme of the heat pump. An adiabatic reciprocating compressor is chosen, in accordance with actual commercial trends [52]. The isentropic efficiency (Equation (7)) is used to model the compressor and the polytropic exponent (n) can be obtained, as shown in Equation (8). The volumetric efficiency (η_v, Equation (9)) is used to set the refrigerant mass flow rate $(\dot{m})$. In Equations (7)–(9), h stands for enthalpy, s entropy, p pressure, v specific volume, vs. swept volume, r pressure ratio, and α relative clearance volume.

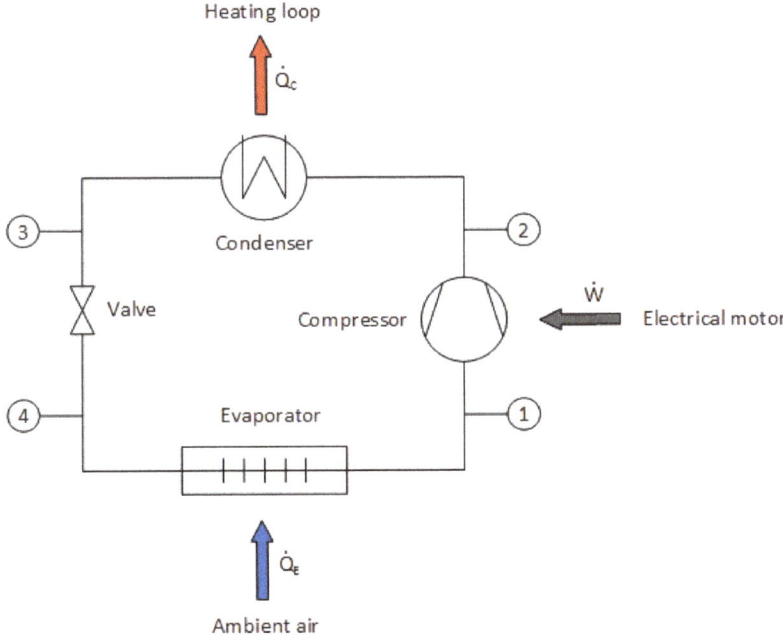

Figure 3. Layout of the heat pump.

The nominal parameters to solve the best efficiency point are:
- Condenser:
  - Heating capacity $(\dot{Q}_c)$: 200 kWth
  - Water conditions: 45 °C at condenser inlet (T_{wi}) and 55 °C at condenser outlet (T_{wo}). A counterflow heat exchanger is assumed, feeding the water to an existing low temperature radiator system.

- Outlet refrigerant conditions: saturated liquid and appropriate approach temperature (ΔT_c) to achieve 5 °C of pinch point (at saturated vapour state). This approach temperature is 13 °C for R290 (propane), which, as will be explained below, has been chosen as the refrigerant of the heat pump.

- Evaporator:
  - Air conditions: 5 °C as ambient wet bulb temperature (taken as evaporator inlet, T_{ai})) and −5 °C as wet bulb temperature at evaporator outlet (T_{ao}). An ambient pressure of 95 kPa is assumed (Madrid exhibits an elevation over the sea level of 655 m).
  - Outlet working conditions: superheating (ΔT_v) of 5 °C
  - Inlet working conditions: approach temperature (ΔT_e) of 10 °C

- Compressor:
  - Speed (N): 1490 rpm
  - Isentropic efficiency (η_s): 75% (includes motor and inverter efficiencies)
  - Relative clearance volume (α): 3%.

$$\eta_s = \frac{h(s_1; p_2) - h_1}{h_2 - h_1} \tag{7}$$

$$r = \left(\frac{p_2}{p_1}\right) = \left(\frac{v_1}{v_2}\right)^n \tag{8}$$

$$\eta_v = \frac{\dot{m}}{\left(\frac{V_s}{v_1}\right)\cdot\left(\frac{N}{60}\right)} = 1 - \alpha\cdot\left(r^{1/n} - 1\right) \tag{9}$$

The condenser is a refrigerant/water counter-flow heat exchanger. Equation (10) shows the energy balance, where $\dot{m}_w$ stands for mass flow rate of water. The approach temperature results in a relationship between both fluids (Equation (11)), where refrigerant outlet temperature is the condensation temperature (refrigerant is saturated liquid at state 3), linked to the condensation pressure (Equation (12)). Enthalpy for water is assessed as enthalpy of saturated liquid at the water temperature. No pressure drop is considered (Equation (13)).

$$\dot{Q}_C = \dot{m}\cdot(h_2 - h_3) = \dot{m}_w\cdot(h_{wo} - h_{wi}) \tag{10}$$

$$\Delta T_c = T_3 - T_{wi} \tag{11}$$

$$T_3 = T_{sat}(p_3) \tag{12}$$

$$p_2 = p_3 \tag{13}$$

The valve is considered to be adiabatic, so, as kinetic and potential energy are neglected, it is modelled as iso-enthalpic (Equation (14)). Besides, the valve is assumed to be a thermostatic expansion valve, so it keeps the superheating at the compressor suction (Equation (15)) constant by acting on the refrigerant mass flow rate.

$$h_3 = h_4 \tag{14}$$

$$\Delta T_v = T_1 - T_4 = constant \tag{15}$$

The evaporator is an air/refrigerant cross-flow heat exchanger. Equation (16) gives the energy balance, where $\dot{m}_a$ stands for air mass flow rate and $(\dot{Q}_E)$ stands for the rate of heat transfer. As in the condenser, the approach temperature establishes a relationship between the fluids (Equation (17)), where refrigerant inlet temperature is the evaporation temperature (refrigerant is a liquid-vapor mixture at state 4), linked to the evaporation pressure (Equation (18)). No pressure drop is considered (Equation (19)).

$$\dot{Q}_E = \dot{m} \cdot (h_1 - h_4) = \dot{m}_a \cdot (h_{ai} - h_{ao}) \quad (16)$$

$$\Delta T_e = T_{ao} - T_4 \quad (17)$$

$$T_4 = T_{sat}(p_4) \quad (18)$$

$$p_4 = p_1 \quad (19)$$

The main performance parameters of the heat pump in its nominal point (BEP) are the heating capacity (useful heat released in the condenser), previously defined, the compressor consumption ($\dot{W}$, Equation (20)), and the COP (Equation (21)). In the design point, all of them are instantaneous values; later on, in the off-design operation, they will be redefined as seasonal parameters, i.e., both the power rates ($\dot{Q}_C$ and $\dot{W}$) and the instantaneous COP will be time-integrated.

$$\dot{W} = \dot{m} \cdot (h_2 - h_1) \quad (20)$$

$$COP = \frac{\dot{Q}_C}{\dot{W}} \quad (21)$$

The refrigerant must comply with regulations about the global warming potential (GWP) and ozone depletion potential (ODP) [11] of the European Union. Accordingly, R-290 (propane) is chosen as fluid valid for long term, due to the fact that it is a natural refrigerant with null ODP and a GWP value of 3. On the other hand, it is included in the A3 class of the flammability safety classification (ASHRAE), thus considered highly flammable. Figure 4 shows the pressure-enthalpy diagram for R-290, where the assumptions previously stated for the best efficiency point are represented. This diagram shows that the compressor discharge temperature is not too high and the de-superheating interval is small (around 20% of the overall heating capacity). Both of the values are in accordance with the fact that domestic hot water is heated by using other procedures.

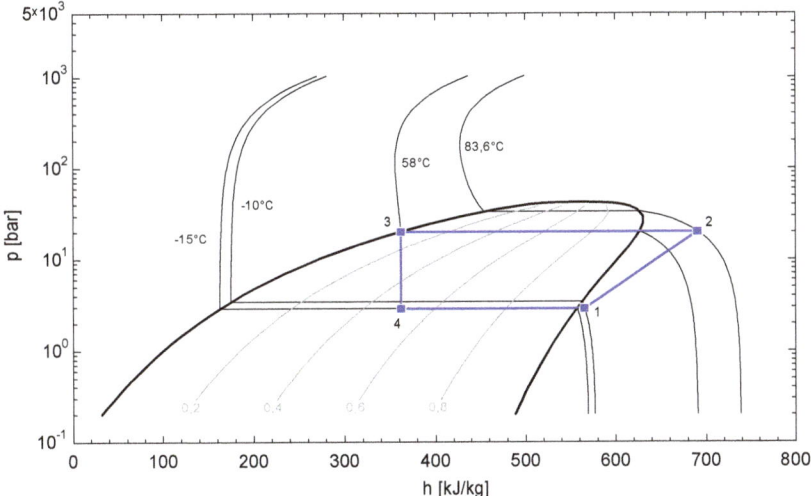

Figure 4. P-h diagram for the design point.

The main parameters of the heat pump are fixed once the design point has been solved. Table 4 summarises these values. At this point (inlet air temperature of 5 °C and outlet water temperature of

55 °C, usually known as A5/W55), the heating capacity is set as 200 kWth, the compressor consumption is found to be 76.15 kWe, and then the COP results in 2.626.

Table 4. Main parameters of the heat pump calculated in the design point and set as constant in off-design operation.

Heat Pump Parameter	
Swept volume, vs. (cm^3/rev)	4558
Polytropic exponent, n (-)	1.075
Condenser approach, ΔT_c (°C)	13
Evaporator approach, ΔT_e (°C)	10
Temperature drop at air, $T_{ai} - T_{ao}$ (°C)	10
Superheating at compressor suction, ΔT_v (°C)	5
Water inlet temperature, T_{wi} (°C)	45
Water outlet temperature, T_{wo} (°C)	55

The modelling of the BEP and the off-design operation have been carried out for different purposes. On one hand, the model of the BEP is developed to define the heat pump parameters that determine its size (listed in Table 4). On the other hand, the off-design model aims to obtain the performance of the heat pump as a function of both the ambient temperature and the heating load, in order to work out the seasonal performance of the device. The input variables are the ambient wet bulb temperature and the rotational speed of the compressor. Equations (8)–(21) are now solved using the parameters that are given in Table 4. It should be noted that the isentropic efficiency of the compressor is replaced by the polytropic relationship (Equation (8)) and the mass flow rates of refrigerant, air, and water are now unknown. Approach temperatures in the heat exchangers have been fixed, while assuming that the number of transfer units are high enough to work in the asymptotic range of effectiveness. The main functions of the off-design model are listed in Equations (22) and (23), which lead to Equation (24). Another limit should be imposed: the maximum driving power of the motor, being assumed as 1.5 times the power consumed at BEP.

$$\dot{Q}_C = \dot{Q}_C(T_{ai}, N) \tag{22}$$

$$COP = COP(T_{ai}, N) \tag{23}$$

$$\dot{W} = \frac{\dot{Q}_C(T_{ai}, N)}{COP(T_{ai}, N)} = \dot{W}(T_{ai}, N) \tag{24}$$

2.3. Coupling of the Demand and Heat Pump Models

Once the hourly heating demand (Equation (6)) and the heat pump performance (Equations (22) and (24)) are determined, the map of the device coupled to the demand can be obtained, producing a diagram similar to the one that is shown in Figure 5. In this chart, the heating demand only considers the linear regression curve for the sake of clarity. At the nominal rotational speed, a positive slope line determines the heating capacity of the heat pump for each temperature. This line cuts to the heating load line in one point. To modulate the response of the heat pump, the rotational speed is varied, therefore sweeping the operation zone and cutting to the load curve in a large range. For temperatures out of that range, the back-up system operates (if the load is higher than the heat pump capacity) or the machine is operated in on/off mode, in order to adapt the excess of capacity to the low demand.

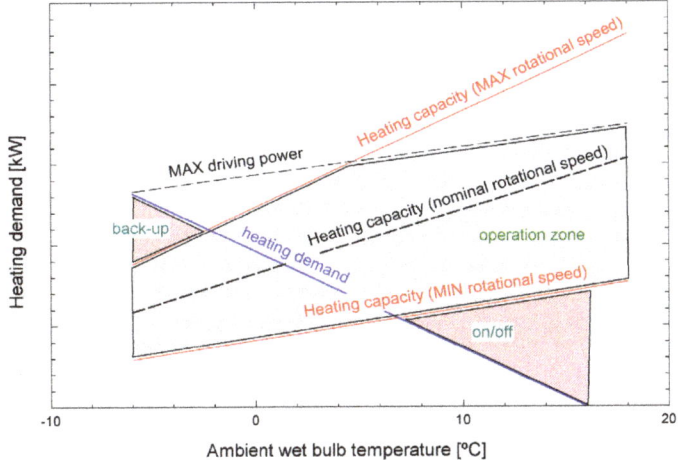

Figure 5. Coupling between demand and performance of the heat pump.

Thus, the consumption of the heat pump ($\dot{W}_j$) and the back-up system ($\dot{F}_j^{bkp}$) (if any) to meet the demand are calculated, for each operation hour, while using the functions that are given in Equation (25a–d). In Equation (25c), *bkp* stands for the back-up system, assumed as a natural gas condensing boiler with a constant efficiency (η_{bkp}) value of 95% on higher heating value basis. In the same equation, $\dot{F}$ stands for the natural gas consumption (again based on HHV) of the back-up system. Equation (25d) determines the power of the back-up boiler ($\dot{Q}_{bkp}$). In the on/off operation range, the demand is covered by the heat pump working at the minimum rotational speed, as it is derived from the algorithm described in Equation (25). Defrosting cycles are neglected, due to the usually low air moisture content in Madrid [53].

$$\dot{Q}_{C,j}^{max} = max\left\{\dot{Q}_C(T_{ai,j}, N_{max}), COP(T_{ai,j}, N_{max}) \cdot 1.5 \cdot \dot{W}\left(T_{ai,j}^{design}, N_{design}\right)\right\} \quad (25a)$$

$$\dot{W}_j = \begin{cases} \dot{Q}_{c,j}/COP(T_{ai,j}, N_j) & \text{if } D_j \leq \dot{Q}_{C,j}^{max} \\ \dot{Q}_{C,j}^{max}/COP(T_{ai,j}, N_j) & \text{otherwise} \end{cases} \quad (25b)$$

$$\dot{F}_j^{bkp} = \begin{cases} 0 & \text{if } D_j \leq \dot{Q}_{C,j}^{max} \\ \left(D_j - \dot{Q}_{C,j}^{max}\right)/\eta_{bkp} & \text{otherwise} \end{cases} \quad (25c)$$

$$\dot{Q}_{bkp} = max\left\{\dot{F}_j^{bkp}\right\} \quad (25d)$$

Some seasonal performance indexes have been defined: heating seasonal performance factor (*HSPF*, Equation (26)), CO_2 avoided ratio (*AVCO2*, Equation (27)) and renewable input to heating demand ratio (*R2H*, Equation (28)). Numerical coefficients that are employed in Equations (27) and (28a) make it possible to consider the environmental impact of the electricity coming from the grid to drive the heat pump. They have been taken from [54], according to the Spanish energy sector. Equation (28b) comes from the current EU regulation regarding the support to heat pumps [10].

$$HSPF = \frac{\sum_{j=1}^{4368} \dot{W}_j \cdot COP(T_{ai,j}, N_j)}{\sum_{j=1}^{4368} \dot{W}_j} \quad (26)$$

$$AVCO2 = \frac{\left(\frac{\sum_{j=1}^{4368} D_j}{\eta_{bkp}}\right) \cdot 0.252 - \left[0.331 \cdot \sum_{j=1}^{4368} \dot{W}_j + 0.252 \cdot \sum_{j=1}^{4368} \dot{F}_j^{bkp}\right]}{\sum_{j=1}^{4368} D_j} \quad (27)$$

$$R2H = \frac{\sum_{j=1}^{4368} RES_j + 0.414 \cdot \sum_{j=1}^{4368} \dot{W}_j}{\sum_{j=1}^{4368} D_j} \quad (28a)$$

$$RES_j = \begin{cases} \dot{Q}_{C,j} \cdot \left(1 - \frac{1}{HSPF}\right) & \text{if } HSPF \geq 2.5275 \\ 0 & \text{otherwise} \end{cases} \quad (28b)$$

2.4. Economic Model

The main indicator to assess the economic feasibility of this kind of device is the levelised cost of heating (LCOH), which integrates both the investment and operating costs [55]. In this research, two LCOH have been calculated: one referred to the whole heating demand ($LCOH_{DB}[€/MWh]$, (Equation (29a)) and another referred to the whole heated area ($LCOH_{AB}[€/m^2]$ (Equation (29b)). In Equation (29a), INV stands for investment, C for annual cost, superscript M refers to maintenance, W to power consumption, F to fuel consumption, subscript 0 to costs in year zero, CRF to the capital recovery factor (Equation (29c)), CELF stands for the constant escalation levelisation factor (Equation (29d)). In Equation (29c–e), r_x represents the nominal escalation rate of the item x, $wacc$ the weighted average capital cost, and N_y is the life span of the project.

$$LCOH_{DB} = \frac{(INV_{HP} + INV_{bkp}) \cdot CRF + C_0^W \cdot CELF^W + C_0^F \cdot CELF^F + C_0^M \cdot CELF^M}{\sum_{j=1}^{4368} D_j} \quad (29a)$$

$$LCOH_{AB} = LCOH_{DB} \cdot \left[\frac{\sum_{j=1}^{4368} D_j}{A}\right] \quad (29b)$$

$$CRF = \frac{wacc \cdot (1 + wacc)^{N_y}}{(1 + wacc)^{N_y} - 1} \quad (29c)$$

$$CELF_x = \left[\frac{k_x \cdot \left(1 - k_x^{N_y}\right)}{1 - k_x}\right] \cdot CRF \quad (29d)$$

$$k_x = \frac{1 + r_x}{1 + wacc} \quad (29e)$$

The investment for the heat pump (inverter model with 200 kWth of heating capacity) has been taken as 24,753 € [52], and a scale law has been fit for the investment of the back-up boiler (Equation (30)).

$$INV_{bkp}[€] = 1087.6 \cdot \dot{Q}_{bkp}^{0.506} \, [kW_{th}] \quad (30)$$

For consumptions with installed power higher than 15 kWe, the cost of electricity includes a term for maximum consumed power in a year and a term for annual consumed energy. Moreover, this consumption considers hourly discrimination in three periods (P1, P2, and P3), according to Table 5.

Table 5. Electrical tariff [56].

Period	Power Term (€/kW) [1]	Energy Term (€/MWh)
P3: 0.00 to 8.00	16.7803	85.3
P2: 8.00 to 18.00	25.1704	114.1
P1: 18.00 to 22.00	41.9507	127.1
P2: 22.00 to 24.00	25.1704	114.1

[1] The cost associated to the power term is derived from multiplying the tariff by the maximum power consumed along the year at each period.

For comparison purposes, two additional scenarios have been considered: a decentralised system and a centralised one, both using only condensing natural gas boilers with the same efficiency as the back-up boiler (95% based on HHV). In the de-centralised boiler case, 60 single dwellings of 100 m² are considered, each employing a natural gas boiler with an investment of 800 € and maintenance costs of 150 €/year. In the centralised boiler scenario, a larger boiler is used to meet the overall demand, being given its investment by Equation (30) and a maintenance cost assumed of 2000 €/year.

All of the costs include taxes. Regarding the natural gas costs, two tariffs have been selected, depending on the annual consumption. Accordingly, for large consumptions (centralised cases in both heat pump with back-up and boiler alone) the tariff is 971.64 €/year plus 40.66 €/MWh (based on HHV), whereas for small consumptions (decentralised boilers) the tariff is 115.98 €/year plus 42.4 €/MWh (again based on HHV) [57].

Maintenance cost has been set to 2000 €/year for the proposed system (heat pump supported by a boiler).

Table 6 shows the economic parameters used to calculate the levelised costs.

Table 6. Assumed economic parameters.

Parameter	Value
Weighted average capital cost, $wacc$ (%)	0
Nominal rate of power, r_W (%)	5
Nominal rate of gas, r_F (%)	5
Nominal rate of maintenance, r_M (%)	2.5
Life span, N_y (years)	15

3. Results

3.1. Heat Pump Model

Once the model is completed, Equations (22)–(24) are solved, obtaining, respectively, Equations (31) to (33). For a given rotational speed, the relation between heat capacity and air temperature is linear, and the slope increases with the speed, as it can be seen in Equation (31). Regarding the COP, Equation (32) is obtained, but it does not depend on rotational speed. This result is foreseeable, because the COP is the ratio between heat capacity and compressor consumption, both proportional to the mass flow rate, which is directly dependent on the rotational speed. Combining both Equations (31) and (32), the expression for the compressor consumption is obtained (Equation (33)).

$$\dot{Q}_C = (0.1106 \cdot N + 0.0002) + (0.0053 \cdot N + 0.0003) \cdot T_{ai} \quad (31)$$

$$COP = 2.49499 + 0.02981 \cdot T_{ai} \quad (32)$$

$$\dot{W} = \frac{(0.1106 \cdot N + 0.0002) + (0.0053 \cdot N + 0.0003) \cdot T_{ai}}{2.49499 + 0.02981 \cdot T_{ai}} \quad (33)$$

Figure 6 shows the iso-lines of heating capacity as a function of the rotational speed and the ambient wet bulb temperature. The limitation of maximum electrical power (Equation (25a)) has been taken into account.

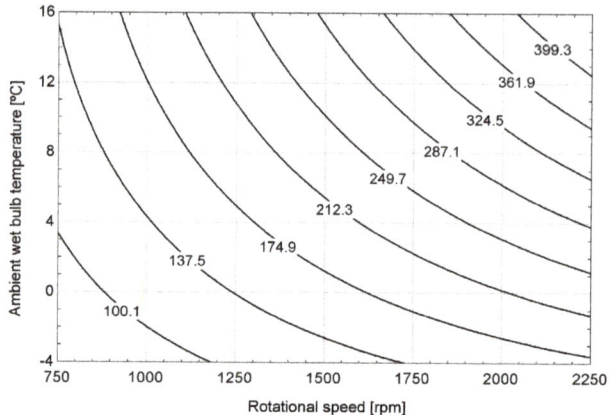

Figure 6. Iso-lines of heating capacity of the ASHPWH.

A comparison with an actual machine has been performed in order to validate the results of the heat pump model. A literature search has been carried out to find an existing device that fits with the characteristics of the modelled one. The selected equipment has been HERA 190-2-2 [58], an air to water heat pump from EUROKLIMAT. A reciprocating compressor using an inverter drives it and it has a design heating capacity of 190 kWth. The refrigerant is R290 as the one selected in this paper. The water outlet temperature of the existing heat pump is 55 °C, as in the model. Figure 7 plots the comparison of the COP, showing a good match, especially at low temperatures. This behavior is due to the fact that the HERA 190-2-2 data are based on dry bulb air temperature, whereas the data from the model are based on the wet bulb temperature, with both temperatures being very similar at low dry bulb temperature.

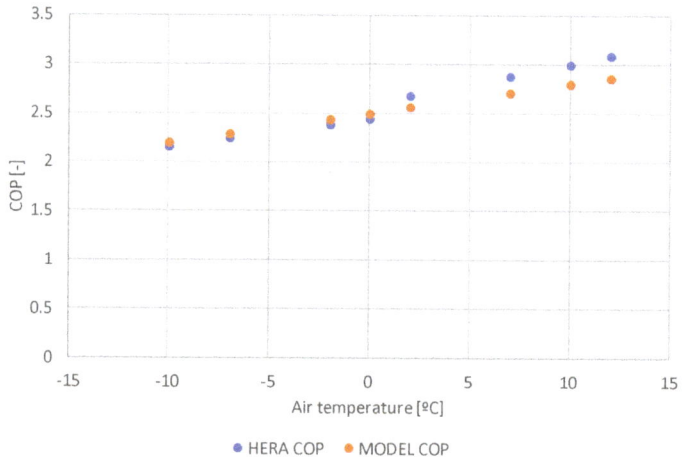

Figure 7. Comparison between the coefficient of performance (COP) obtained with the model and the COP of HERA 190-2-2 from EUROKLIMAT.

3.2. Baseline Case

A block dwelling with an overall heated area of 6000 m² and EPI of 2.18, i.e., built between 1981 and 2007, located in Madrid, has been simulated. It is important to point out that the selected baseline-building-age represents a considerable percentage of vulnerable households in Spain and the 41% of the overall main houses with heating [46]. Figure 2 shows its annual cumulative heating demand profile. Applying the algorithm given by Equation (25) to each winter hour, the results that are shown in Table 7 are obtained. From these values, the performance indexes that are given by Equations (26)–(28) are calculated, and Table 8 summarises the results. Furthermore, Figure 8 shows the contribution to the heating demand of both the heat pump (96%) and the back-up boiler (4%).

Table 7. Energy results in the baseline case.

Parameter	Value
Seasonal heating demand, $\sum_{j=1}^{4368} D_j$ (MWh)	600.833
Heating demand met by heat pump, $\sum_{j=1}^{4368} \dot{W}_j \cdot COP_j$ (MWh)	576.675
Seasonal consumption of heat pump, $\sum_{j=1}^{4368} \dot{W}_j$ (MWh)	223.080
Back-up boiler consumption, $\sum_{j=1}^{4368} \dot{F}_j^{bkp}$ (MWh)	25.430
Size of back-up boiler, $\dot{Q}_{bkp}$ (kW)	180
Renewable energy taken from the ambient air, $\sum_{j=1}^{4368} RES_j$ (MWh)	445.949

Table 8. Performance indexes in the baseline case.

Parameter	Value
Heating seasonal performance factor, HSPF (-)	2.585
Avoided CO_2 emissions, AVCO2 (g CO_2/kWht)	131.7
Renewable to heating demand ratio, R2H (%)	74.22

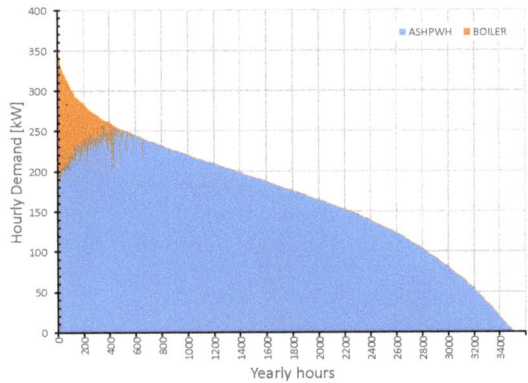

Figure 8. Contribution of heat pump (ASHPWH) and back-up boiler (BOILER) to meet the annual cumulative heating demand profile in the baseline case (block dwellings in Madrid built between 1981 to 2007 with 6000 m² of total heated surface).

At low thermal demands, the heat pump is controlled in on/off mode due to the minimum rotational speed limit. This type of control causes important energy losses when an inverter is not available, but this driving control is expected to reduce them. Accordingly, Figure 9 shows the cloud of points for the thermal demand under the minimum heat capacity line. Once these points are sorted and moved into the ASHPWH contribution to the annual cumulative heating demand profile, they take up the end tale of the distribution (Figure 10), accounting for 8% of the overall demand that is met by the heat pump. Figure 9 also shows that the motor consumption ranges between 31.2 to 49.9 kWe (41 to 65.5% as compared to the consumption at design point). This allows for us to assume that low current peaks will take place in the startups and, moreover, they also will be smoothed by the use of the speed modulation. On the other hand, the ratio of the thermal demand to the minimum heating capacity of the heat pump at each hour determines the time fraction when the heat pump has been working over that hour. Figure 11 shows this ratio, after being sorted. It can be observed that in 58% of the time when the heat pump was working in on/off mode was "on" more than 30 min (hour fraction 0.5), therefore minimizing the transitory effect on the components of the heat pump. In summary, the availability of the inverter is expected to significantly contribute to reducing the energy losses of the on/off operation mode and, consequently, it make sense to neglect it for the scope of the current analysis.

Regarding the costs, Table 9 summarises the results.

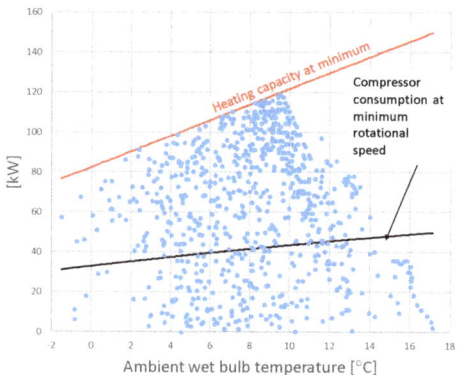

Figure 9. Thermal demand points in on/off operation mode versus the ambient wet bulb temperature. Heating capacity and compressor consumption at minimum rotational speed are also plotted.

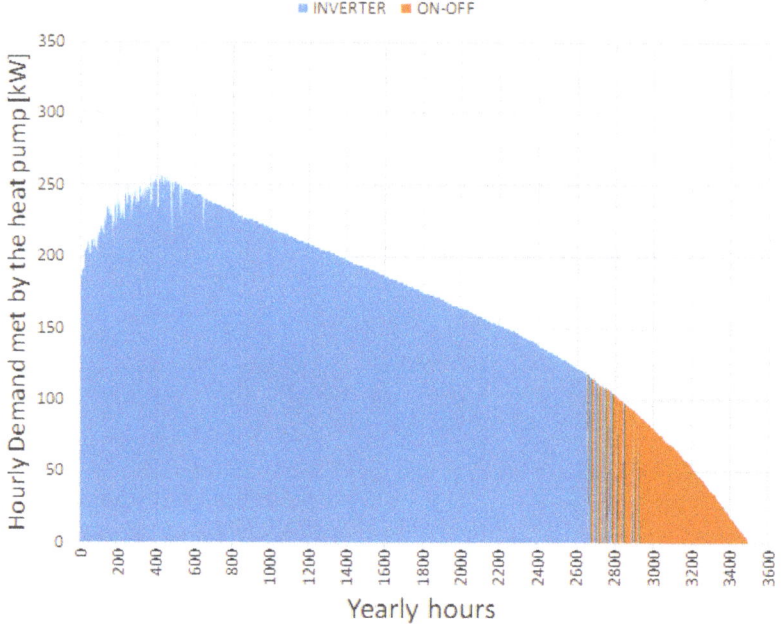

Figure 10. Thermal demand met by the heat pump at inverter control mode and on/off one.

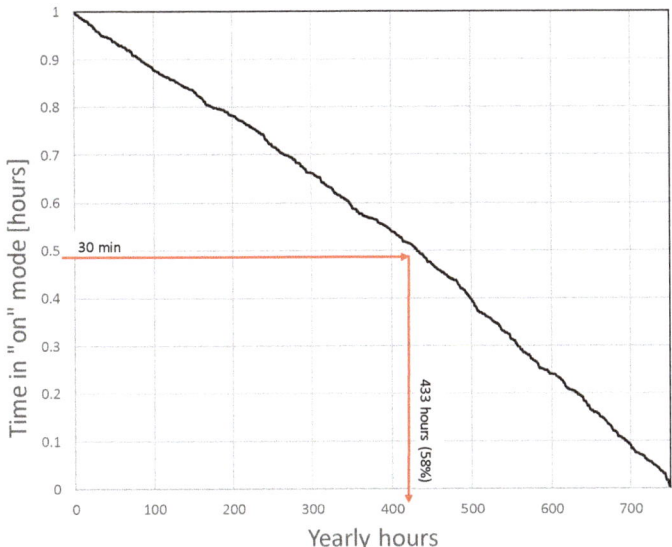

Figure 11. Hour fraction (time in hours) when the heat pump was "on", while working in on/off mode.

Table 9. Levelised costs in the baseline case for heat pump, de-centralised, and centralised boilers.

Levelised Costs	Heat Pump	De-Centralised Boiler	Centralised Boiler
Levelised cost of heating (demand based), $LCOH_{DB}$ (€/MWh)	92.22	108.55	73.54
Levelised cost of heating (area based), $LCOH_{AB}$ (€/m^2)	9.23	10.87	7.36

Figure 12 shows the breakdown of the levelised costs that are given in Table 9. As the demand and the heated area are established, the percentage breakdown is the same for both levelised costs (see Equation (29a,b)). Moreover, Figure 12 points out the predominance of operating costs (especially the energy term, being the maintenance less significant) over investment (heat pump and boiler).

The proposed active measure should be supplemented by the application of a social tariff based on the cost breakdown. This would make it possible to take advantage of the environmental benefits of the proposed system at the same levelised cost of the most economical system, i.e., the centralised boiler. In this case, a discount of 23.75% in the electricity cost (overall cost, including power and energy terms) would reduce the LCOH to 73.54 €/MWh, matching the cost of the centralised boiler. This discount is in accordance with the current social electricity tariff in Spain [59], which ranges from 25% for vulnerable consumers, 40% for severely vulnerable consumers, and up to 100% for consumers at risk of social exclusion. The average discount was 32% when considering the distribution of each cluster in 2019 (respectively, 648,826 and 630,086 households in the first two categories, and 4545 in the third one [60]). Therefore, the required discount to spread the proposed system would be even lower than the one currently applied to vulnerable-households' electricity bill.

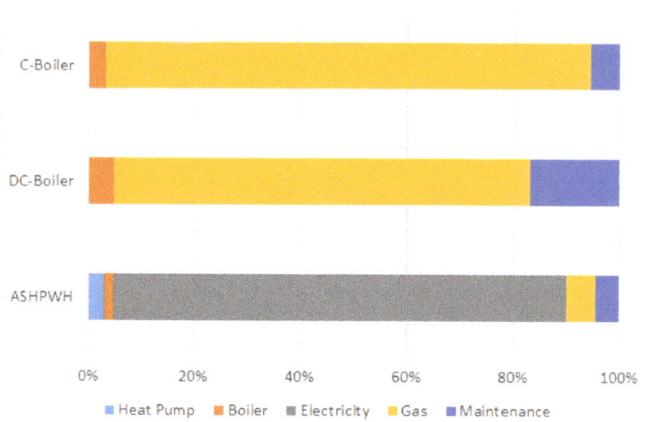

Figure 12. Percentage breakdown of the levelised costs.

3.3. Influence of Energy Retrofitting

The baseline case has been selected to be representative of the middle-energy efficiency level pool of dwellings, with an energy performance assessment of E, close to D (EPI = 2.18). However, in the context of energy poverty, it is usual to find dwellings built before 1980 with worse insulation condition than this middle energy-efficiency-level. In these cases, an energy retrofitting of the dwelling is recommended, which makes it possible to reduce the heating cost per dwelling. To assess this fact, Figure 13 has been obtained, varying EPI and maintaining the levelised cost of heating in demand base with the same value as the baseline case. This condition requires varying the heated area, obtaining a potential fitting curve (Equation (34)). In Figure 13 the average EPI obtained for dwelling blocks in Spain have been represented over such fitting curve, so obtaining their heated area to maintain the same levelised cost than in the baseline case.

$$A = 11,139 \cdot EPI^{-0.789} \tag{34}$$

4. Discussion

The proposed demand model makes it possible to obtain the hourly demand profile of a building, which is essential for working out the instantaneous consumptions of the heating system. The shape of the annual cumulative heating demand curve is consistent with the profiles that were obtained by other simulation tools [40]. This model only requires the location of the building, its useful area and the energy performance certification (EPC). A detailed simulation of the building thermal behaviour is necessary to assess the EPC, which leads to the EPI that is integrated in the proposed model. The fact that no specific additional information about the building is required by the model makes it very easy-to-use and helpful for planning purposes. The flexibility of the model has been used in order to solve a case study representative of vulnerable dwellings.

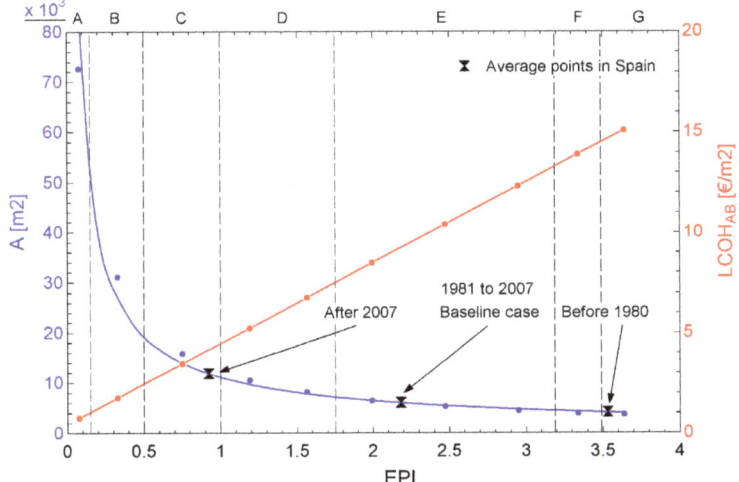

Figure 13. Required heated area (A) for each energy performance index (EPI) to maintain the same $LCOH_{DB}$ than in the baseline case (92.22 €/MWh) and $LCOH_{AB}$ obtained. Energy performance classes (A to G) limits are indicated in dashed lines.

Regarding the heat pump, R290 (propane) has revealed as a suitable refrigerant for this application, and it is also supported by many manufacturers. It is a natural fluid, with zero ODP and very low GWP. The low compressor discharge temperature and the low de-superheating zone in the condenser advise taking this fluid into consideration for the future. Its high flammability is not relevant in the current case, since it is a centralised unit, with roof-top allocation and maintenance performed by professional workers. The speed control in the compressor gives to the heat pump the capability to meet nearly all of the heating demand (96% in the baseline case) with good efficiency. In the baseline case, the heating seasonal performance factor achieved is 2.585, exceeding the minimum that is required to consider the thermal energy taken from the air as renewable energy. Furthermore, the overall renewable energy to meet the heating demand is obtained, taking both the energy supplied by the air and the renewable share in the electricity mix into account (applied to the electricity consumption of the heat pump). Therefore, each kWh-th met with the proposed system avoids the emission of 131.7 g CO_2, whereas 265.3 g CO_2 are emitted if a boiler is used.

Regarding the cost for the baseline case, the proposed technology (heat pump plus boiler back-up) reduces the levelised cost of heating by 15% with respect to the de-centralised boiler. However, the proposed system increases the cost by 25% with respect to the centralised boiler. On the other hand, with centralised gas neither renewables sources are employed nor carbon dioxide emissions are avoided. The cost breakdown reveals that the energy cost, which facilitates the integration of subsidy policies to the operation, is the most important contribution to the overall cost. Therefore, a discount of 23.75% in electricity bill (comparable with the current social-tariff average-discount) would be enough to equalise the levelised cost of heating of the proposed system with the centralised boiler (the most economical scenario). Furthermore, the innovative and social aspect of this system makes it possible to integrate funding by non-profit organisations, energy companies, Public Administration, or even the actual consumers (depending on their vulnerability level). These subsidies to the proposed system would be endorsed due to its excellent environmental performance, balancing this way de-carbonisation with energy affordability.

Finally, the effect of energy retrofitting has been investigated. The results show that, if the energy performance assessment is improved from E to C due to a retrofit, the same heat pump would be able to meet the demand of 160 dwellings of 100 m^2 instead of 60 of the baseline case. Regarding the heating cost per square meter, the cost drops 4 €/m² for every EPI unit reduction. Accordingly, in the same example, the levelised cost of heating decreases 500 € for a dwelling of 100 m² (with a baseline cost of 848 €).

5. Conclusions

The feasibility of heating a block dwelling in Madrid by an air-source heat-pump water heater has been investigated in the framework of energy poverty research. The implementation of the heat pump is planned as a retrofit of the heating system. Therefore, an existing system that is based on radiators is assumed, and the operation temperatures of the heat pump are adapted to this configuration. A global methodology has been used to forecast the hourly heating demand while using an expansion method that is based on the Spanish regulation. The performance map has been obtained after selecting an eco-friendly refrigerant (R290), and considering a speed variation control over the compressor. This control makes it possible to meet nearly the whole demand with a high efficiency. Finally, the demand has been coupled to the performance of the heat pump.

When considering environmental and efficiency indicators, the obtained results show an excellent behaviour of the heat-pump as compared with the classical solution of the natural gas centralised boiler system. However, the levelised cost of heating is 25% higher than the centralised boiler, due to the low prices of gas for high consumption volumes. In this sense, a discount in electricity bill for vulnerable households might be taken into account, considering both social and environmental benefits, to equalise the energy costs of this technology with the one of a centralised gas boiler. The situation is reversed in the case of comparing the heat pump with the de-centralised boiler solution, with the heating cost of the heat pump being 15% lower than the boiler one. In all the scenarios, the main contribution to the overall cost is the operation, especially the energy cost.

Therefore, the heat pump has revealed as an efficient and sustainable system to tackle energy poverty in a typical case of vulnerable households, such as the building blocks analysed, especially when compared with a de-centralised boiler heating system. In any case, a new electricity tariff frame that reduces the operation costs of heat pumps with respect to the centralised boiler solution would be highly recommended.

In future works, the model will be expanded to summer season (cooling demand), taking advantage of the reversible ability of heat pumps. In the context of energy poverty, this is a new trend, especially in countries in Southern Europe. Other technologies, such as thermally driven heat pumps (both absorption and internal combustion engine), might be also considered to be alternatives. Moreover, the application to the different climatic zones of Spain will be carried out and policy implications will be pointed out.

Author Contributions: Conceptualization, J.I.L. and R.B.; methodology, E.A.; software, I.P.; validation, J.C.R. and E.C.; formal analysis, J.I.L. and R.B.; investigation, I.P.; resources, E.A.; writing—original draft preparation, J.I.L.; writing—review and editing, R.B. and E.A.; supervision, J.C.R.; project administration, J.C.R.; funding acquisition, E.C. All authors have read and agreed to the published version of the manuscript.

Funding: This research was funded by Chair of Energy and Poverty of Comillas Pontifical University.

Conflicts of Interest: The authors declare no conflict of interest.

References

1. Thomson, H.; Bouzarovski, S. Addressing Energy Poverty in the European Union: State of Play and Action. EU Energy Poverty Observatory, European Commission. 2018. Available online: https://www.energypoverty.eu/sites/default/files/downloads/publications/19-05/paneureport2018_updated2019.pdf (accessed on 9 April 2020).
2. EU Statistics on Income and Living Conditions 2017. Available online: https://ec.europa.eu/eurostat/web/microdata/european-union-statistics-on-income-and-living-conditions (accessed on 9 April 2020).
3. Romero, J.C.; Linares, P.; López, X. The policy implications of energy poverty indicators. *Energy Policy* **2018**, *115*, 98–108. [CrossRef]

4. Ministry for Industry and Ecological Transition, Spanish Strategy to Tackle Energy Poverty. 2019. Available online: https://www.miteco.gob.es/es/ministerio/planes-estrategias/estrategia-pobreza-energetica/actualizaciondeindicadorespobrezaenergetica2019_tcm30-502983.pdf (accessed on 15 April 2020).
5. Dobbins, A.; Fuso Nerini, F.; Deane, P.; Pye, S. Strengthening the EU response to energy poverty. *Nat. Energy* **2019**, *4*, 2–5. [CrossRef]
6. Boardman, B. Fuel poverty synthesis: Lessons learnt, actions needed. *Energy Policy* **2012**, *49*, 143–148.
7. Marmot Review Team. *The Health Impacts of Cold Homes and Fuel Poverty*; The Baring Foundation: London, UK, 2011.
8. Cruz Roja Española. La Vulnerabilidad asociada al ámbito de la vivienda y pobreza energética en la población atendida por Cruz Roja. Available online: https://www.cruzroja.es/principal/documents/1789243/2038966/Informe_Cruz_Roja_Boletin_sobre_la_vulnerabilidad_social_N17_Vivienda_Pobreza_Energ%C3%A9tica.pdf/59045195-3960-d9a5-d632-7a92664df97a (accessed on 26 May 2020).
9. Vedavarz, A.; Kumar, S.; Hussain, M.I. *HVAC: Handbook of Heating, Ventilation and Air Conditioning for Design and Implementation*; Industrial Press Inc.: New York, NY, USA, 2007.
10. EUR-Lex. Directive 2009/28/EC of the European Parliament and the Council of 23 April 2009 on the Promotion of the Use of Energy from Renewable Sources and Amending and Subsequently Repealing Directives 2001/77/EC and 2003/30/EC. 2009. Available online: https://eur-lex.europa.eu/legal-content/EN/ALL/?uri=CELEX%3A32009L0028 (accessed on 26 May 2020).
11. EUR-Lex. Regulation (EU) No 517/2014 of the European Parliament and of the Council of 16 April 2014 on Fluorinated Greenhouse Gases and Repealing Regulation (EC) No 842/2006. 2014. Available online: https://eur-lex.europa.eu/eli/reg/2014/517/oj (accessed on 26 May 2020).
12. Building Technical Code (Release 2019). Available online: https://www.codigotecnico.org/images/stories/pdf/ahorroEnergia/DBHE.pdf (accessed on 9 April 2020).
13. Underwood, C.P.; Royapoor, M.; Sturm, B. Parametric modelling of domestic air-source heat pumps. *Energy Build.* **2017**, *139*, 578–589. [CrossRef]
14. Lohani, S.P.; Schmidt, D. Comparison of energy and exergy analysis of fossil plant, ground and air source heat pump building heating system. *Renew. Energy* **2010**, *35*, 1275–1282. [CrossRef]
15. Tangwe, S.; Simon, M.; Meyer, E.L.; Mwampheli, S.; Makaka, G. Performance optimization of an air source heat pump water heater using mathematical modelling. *J. Energy S. Afr.* **2015**, *26*, 96–105. [CrossRef]
16. Vieira, A.S.; Stewart, R.A.; Beal, C.D. Air source heat pump water heater in residential buildings in Australia: Identification of key performance parameters. *Energy Build.* **2015**, *91*, 148–162. [CrossRef]
17. Çengel, Y. *Heat Transfer: A Practical Approach*; McGraw-Hill Education: Boston, MA, USA, 2002.
18. Fardoun, F.; Ibrahim, O.; Zoughaib, A. Quasi-steady state modeling of an air source heat pump water heater. *Energy Procedia* **2011**, *6*, 325–330. [CrossRef]
19. Patnode, A.M. Simulation and Performance Evaluation of Parabolic trough Solar Power Plants. Master's Thesis, University of Wisconsin-Madison, Madison, WI, USA, 2006.
20. Bourke, G.; Bansal, P. Energy consumption modeling of air source electric heat pump heaters. *Appl. Therm. Eng.* **2010**, *30*, 1769–1774. [CrossRef]
21. Dong, L.; Li, Y.; Mu, B.; Xiao, Y. Self-optimizing control of air-source heat pump with multivariable extremum seeking. *Appl. Therm. Eng.* **2015**, *84*, 180–195. [CrossRef]
22. Cabrol, L.; Rowley, P. Towards lo carbon homes—A simulation analysis of building-integrated air-source heat pump systems. *Energy Build.* **2012**, *48*, 127–136. [CrossRef]
23. Ibrahim, O.; Fardoun, F.; Younes, R.; Louahlia-Gualous, H. Air source heat pump water heater: Dynamic modelling, optimal energy management and mini-tubes condensers. *Energy* **2014**, *64*, 1102–1116. [CrossRef]
24. Xu, D.; Qu, M. Energy, environmental, and economic evaluation of a CCHP system for a data center based on operational data. *Energy Build.* **2013**, *67*, 176–186. [CrossRef]
25. Gadd, H.; Werner, S. Heat load patterns in district heating substations. *Appl. Energy* **2013**, *108*, 176–183. [CrossRef]
26. Bacher, P.; Madsen, H.; Nielsen, H.A.; Perers, B. Short-term heat load forecasting for single-family houses. *Energy Build.* **2013**, *65*, 101–112. [CrossRef]
27. Noussan, M.; Abdin, G.C.; Poggio, A.; Roberto, R. Biomass-fired CHP and heat storage system simulations in existing district heating systems. *Appl. Therm. Eng.* **2014**, *71*, 729–735. [CrossRef]
28. EnergyPlus. Available online: energyplus.net (accessed on 9 April 2020).

29. Wood, S.R.; Rowley, P.N. A techno-economic analysis of small-scale, biomass fuelled combined heat and power for community housing. *Biomass Bioenergy* **2011**, *35*, 3849–3858. [CrossRef]
30. Michopoulos, A.; Skoulou, V.; Voulgari, V.; Tsikaloudaki, A.; Kyriakis, N.A. The exploitation of biomass for building space heating in Greece: Energy, environmental and economic considerations. *Energy Convers. Manag.* **2014**, *78*, 276–285. [CrossRef]
31. Pedersen, L.; Stang, J.; Ulseth, R. Load prediction method for heat and electricity demand in buildings for the purpose of planning for mixed energy distribution systems. *Energy Build.* **2008**, *40*, 1124–1134. [CrossRef]
32. Yun, K.; Luck, R.; Mago, P.J.; Cho, H. Building hourly thermal load prediction using an indexed ARX model. *Energy Build.* **2012**, *54*, 225–233. [CrossRef]
33. Powell, K.M.; Sriprasad, A.; Cole, W.J.; Edgar, T.F. Heating, cooling, and electrical load forecasting for large-scale district energy system. *Energy* **2014**, *74*, 877–885. [CrossRef]
34. Büyükalaca, O.; Bulut, H.; Yılmaz, T. Analysis of variable-base heating and cooling degree-days for Turkey. *Appl. Energy* **2001**, *69*, 269–283. [CrossRef]
35. Martinaitis, V.; Bieksa, D.; Miseviciute, V. Degree-days for the exergy analysis of buildings. *Energy Build.* **2010**, *42*, 1063–1069. [CrossRef]
36. Layberry, R.L. Analysis of errors in degree days for building energy analysis using meteorological office weather station data. *Build. Serv. Eng. Resour. Technol.* **2009**, *30*, 79–86. [CrossRef]
37. Carlos, J.S.; Nepomuceno, M.C.S. A simple methodology to predict heating load at an early design stage of dwellings. *Energy Build.* **2012**, *55*, 198–207. [CrossRef]
38. Sánchez, F.; Álvarez, S. Modelling microclimate in urban environments and assessing its influence on the performance of surrounding buildings. *Energy Build.* **2004**, *36*, 403–414.
39. Sánchez de la Flor, F.J.; Álvarez, S.; Molina, J.L.; González, R. Climatic zoning and its application to Spanish building energy performance regulations. *Energy Build.* **2008**, *40*, 1984–1990. [CrossRef]
40. Uris, M.; Linares, J.I.; Arenas, E. Size optimization of a biomass-fired cogeneration plant CHP/CCHP (Combined heat and power/Combined heat, cooling and power) based on Organic Rankine Cycle for a district network in Spain. *Energy* **2015**, *88*, 935–945. [CrossRef]
41. Ministry of Development, Ministry of Industry, Tourism and Trade (IDAE, Institute for Diversification and Saving of Energy). Energy Performance Scale. Existing Buildings. 2011. Available online: https://www.idae.es/uploads/documentos/documentos_11261_EscalaCalifEnerg_EdifExistentes_2011_accesible_c762988d.pdf (accessed on 9 April 2020).
42. Besagni, G.; Borgarello, M. The socio-demographic and geographical dimensions of fuel poverty in Italy. *Energy Res. Soc. Sci.* **2019**, *49*, 192–203. [CrossRef]
43. Building Technical Code, Basic Document HE (HE-1, Appendix D), April 2009. Available online: https://www.codigotecnico.org/images/stories/pdf/ahorroEnergia/historico/DBHE_200801.pdf (accessed on 9 April 2020).
44. DB-HE (MET Files). Available online: https://www.codigotecnico.org/images/stories/pdf/ahorroEnergia/CTE datosMET_20140418.zip (accessed on 9 April 2020).
45. Ministry of Development, Ministry of Industry, Tourism and Trade (IDAE, Institute for Diversification and Saving of Energy). Energy Performance Scale. 2015. Available online: https://energia.gob.es/desarrollo/EficienciaEnergetica/CertificacionEnergetica/DocumentosReconocidos/normativamodelosutilizacion/20151123-Calificacion-eficiencia-energetica-edificios.pdf (accessed on 9 April 2020).
46. INE (Spanish National Institute of Statistics). CENSUS 2011. 2011. Available online: https://www.ine.es/censos2011_datos/cen11_datos_inicio.htm (accessed on 20 February 2020).
47. Ministry of Development, IDAE (Institute for the Diversification and Saving of Energy), and Ministry of Industry, Trade and Tourism. Estado de la Certificación Energética de los Edificios. 2017. Available online: https://www.certificadosenergeticos.com/wp-content/uploads/2018/12/informe-seguimiento-certificacion-energetica.pdf (accessed on 26 May 2020).
48. Martín-Consuegra, F.; de Frutos, F.; Oteiza, I.; Agustín, H.A. Use of cadastral data to assess urban scale building energy loss. Application to a deprived quarter in Madrid. *Energy Build.* **2018**, *171*, 50–63. [CrossRef]
49. Martín-Consuegra, F.; Hernández-Aja, A.; Oteiza, I.; Alonso, C. Distribución de la pobreza energética en la ciudad de Madrid (España). *EURE* **2019**, *45*, 133–152. [CrossRef]
50. Fabbri, K. Building and fuel poverty, an index to measure fuel poverty: An Italian case study. *Energy* **2015**, *89*, 244–258. [CrossRef]

51. BAXI. Available online: https://www.baxi.co.uk/our-boilers/ (accessed on 12 May 2020).
52. ENERBLUE. Available online: https://enerblue.it/en/products (accessed on 9 April 2020).
53. Tabatabaei, S.A.; Treur, J.; Waumans, E. Comparative evaluation of different computational models for performance of air source heat pumps based on real world data. *Energy Procedia* **2016**, *95*, 459–466. [CrossRef]
54. Ministry of Development, Ministry of Industry, Energy and Trade. CO2 Emission Factors and Pass Coefficients to Primary Energy of Different Final Energy Sources Consumed in the Building Sector at Spain. 2016. Available online: https://energia.gob.es/desarrollo/EficienciaEnergetica/RITE/Reconocidos/Reconocidos/Otros%20documentos/Factores_emision_CO2.pdf (accessed on 9 April 2020).
55. Bejan, A.; Tsatsaronis, G.; Moran, M. *Thermal Design & Optimization*; John Wiley & Sons: New York, NY, USA, 1996.
56. ENDESA. Available online: https://www.endesa.com/es/empresas/luz/tarifa-optima (accessed on 12 December 2019).
57. NATURGY. Available online: https://tarifasgasluz.com/comercializadoras/naturgy/tarifas (accessed on 12 December 2019).
58. EUROKLIMAT. Available online: http://euroklimat.it/download_allegato.php?tabella=12&campo=documento&id=348 (accessed on 12 May 2020).
59. Royal Decree Law 15/2018, 5 October, about Urgent Measures for Energy Transition and Consumer Protection. Available online: https://www.boe.es/diario_boe/txt.php?id=BOE-A-2018-13593 (accessed on 12 May 2020).
60. Ministry for Ecological Transition and Demographic Challenge. Data Supplied through the Information and Attention to the Citizen Channel. Available online: https://www.miteco.gob.es/es/ministerio/servicios/informacion/informacion-y-atencion-al-ciudadano/default.aspx (accessed on 14 May 2020).

© 2020 by the authors. Licensee MDPI, Basel, Switzerland. This article is an open access article distributed under the terms and conditions of the Creative Commons Attribution (CC BY) license (http://creativecommons.org/licenses/by/4.0/).

Article

Applicability of Swaging as an Alternative for the Fabrication of Accident-Tolerant Fuel Cladding

Dae Yun Kim [1], You Na Lee [1], Joon Han Kim [2], Yonghee Kim [3,*] and Young Soo Yoon [1,*]

1. Department of Materials Science and Engineering, Gachon University, Gyeonggi-do 13120, Korea; ct1352@gc.gachon.ac.kr (D.Y.K.); ynl87@gc.gachon.ac.kr (Y.N.L.)
2. SFR Nuclear Fuel Development Division, Korea Atomic Energy Research Institute, Daejeon 34507, Korea; junhkim@kaeri.re.kr
3. Department of Nuclear and Quantum Engineering, Korea Advanced Institute of Science and Technology, Daejeon 34141, Korea
* Correspondence: yongheekim@kaist.ac.kr (Y.K.); benedicto@gachon.ac.kr (Y.S.Y.)

Received: 25 May 2020; Accepted: 17 June 2020; Published: 19 June 2020

Abstract: We suggest an alternative to conventional coating methods for accident-tolerant fuel (ATF) cladding. A Zircaloy-4 tube was inserted into metal tubes of different materials and the inserted tubes were subjected to physical force at room temperature. The manufactured tube exhibited a pseudo-single tube (PST) structure and had higher thermal stability than a Zircaloy-4 tube. Optical microscopy and scanning electron microscopy images showed that the PST had a uniform and well-bonded interface structure, i.e., no gaps or voids were found at the interface between the inner and outer tubes. Energy-dispersive X-ray spectroscopy analysis confirmed that the metal components did not interdiffuse at the interface of the PST, even after being kept at 600 and 900 °C for 1 h and rapidly cooled to room temperature. Unlike pure Zircaloy-4 tubes, Zircaloy-4/stainless use steel (SUS) 316 PST did not show significant structural collapse, even after being stored at 1200 °C for 1 h. Based on these results, if a PST was fabricated using a Zircaloy-4 tube thinner than the Zircaloy-4 tube used in this study and an outer tube of micron-scale thickness, swaging may be a feasible alternative to Zircaloy-4-based ATF cladding.

Keywords: alternative process; non-coating method; room-temperature swaging; pseudo-single tube (PST); accident-tolerant fuel (ATF) cladding

---

1. Introduction

Nuclear fuel claddings are crucial core materials in nuclear power plants. They effectively transfer the heat generated by the fission reaction to the coolant, while preventing the fuel and fission products from leaking into the coolant. Therefore, selecting cladding materials with high corrosion resistance, suitable mechanical properties, and the ability to withstand high pressures, temperatures, and irradiation is vital. Zirconium alloys have shown high potential as cladding materials, owing to their low neutron absorption rate, high corrosion resistance, and stable mechanical properties under the operating conditions of the reactor [1–4]. However, zirconium alloys were found to lack sufficient physical and chemical stability in the Fukushima nuclear power plant accident, highlighting the necessity for research on the development of accident-tolerant fuel (ATF) claddings with improved physicochemical stability. Several alternative ATF claddings have been developed in and outside of Korea, including the M5 physical vapor deposition (PVD)-coated cladding by AREVA [5], the Fe-Cr-Al alloy by ORNL [6], the SiC/SiCf composite by MIT [7], and the Cr laser coating by KAERI [8,9]. However, the ATF cladding fabrication techniques described above have the following disadvantages: it is difficult to achieve uniform surface modifications, such as coatings, on long fuel rods, and issues such as peeling can occur due to thermal shock [10]; FeCrAl alloys experience neutron absorption and tritium emission problems [11]; SiC/SiC$_f$ composite claddings consist of ceramics and are therefore exceedingly unstable in terms of fatigue behavior; and, finally, the abovementioned claddings are difficult to mass-produce, and also incur high production costs [12,13]. To overcome these shortcomings, we propose swaging as an alternative method for the production of ATF claddings. Swaging is a simple process and can be carried out at room temperature. In general, the term "swaging" refers to the process of creating a tube with an outer diameter of a desired size by applying a physical force toward the center of a tube with a larger outer diameter. In this study, swaging indicates the process whereby a tube is inserted into another tube that has an inner diameter greater than the outer diameter of the inserted tube; subsequently, these tubes are joined by applying compressive and tensile stresses toward the center and along the lengths to form a pseudo-single tube (PST) without gaps. If a metal or alloy material with physicochemical properties suitable for nuclear power plant operations is produced in the form of a tube into which Zircaloy-4 can be inserted, there is a possibility that a PST of a length of several meters (more than 1 m) can be easily manufactured using swaging. This process requires a 4 t/cm^2 load rolling press, and the finished PST has an increased length and a reduced outer diameter compared with the starting tube. The entire swaging process is performed at room temperature; hence, there are no heat-induced formations of microstructures or phase deformations during the process. A heat treatment test confirmed that the interface of the PST remained thermally stable at high temperatures. Furthermore, with swaging, surface modification is possible without length constraints and the results of heat treatment suggest that the PST is suitable for use as an ATF cladding [14–16]. The biggest advantage of swaging compared to the coating method is that uniform cladding can be manufactured in a short time at low cost. This advantage can be increased as the size of cladding increases, which leads to the expectation of mass productivity. The coating method may be more advantageous than swaging to set the optimum point of the cladding thickness but there may be many shortcomings in terms of practical use [17]. When cladding is formed over the entire area of a metric cylindrical tube, it is necessary to verify whether the optimal thickness can be formed uniformly. This is important because it is directly related to the safety of nuclear energy. The produced PSTs can also be applied in other industries in the form of metal tubes, which can be manufactured with the desired physical and chemical properties, depending on the choice of materials and the thickness of the outer and inner tubes.

Although in this study, a sample with an ATF geometry (cladding length and thickness) that is directly applicable to nuclear operations has not been used, a 90 cm-long PST was constructed of stainless use steel (SUS) 316 (outside) and Zircaloy-4 (inside) and its high-temperature stability was confirmed. Based on this, we suggest that swaging can be an alternative process for the fabrication of ATF claddings. For this purpose, the interface between the two metals of a PST tube was analyzed before and after high-temperature heat treatment by optical microscopy (OM) and scanning electron microscopy (SEM). Furthermore, Zircaloy-4 cladding and Zircaloy-4 with SUS 316 on the surface, i.e., PST, were exposed to high temperatures for the same durations and their physicochemical stabilities were compared. The core of this study is to suggest a simple and practical process method that can compensate for the disadvantages of the existing coating method. The most frequently mentioned cladding materials are Cr or Cr-based alloys and studies to optimize them are still incomplete. For the swaging process, a cladding tube manufactured in the form of a pipe is required and the difficulty of its manufacture serves as a barrier to the progress of this study [18]. In this situation, we have conducted prior studies on the applicability of the swaging process for ATF cladding by focusing on the process method using SUS 316, which has excellent oxidation resistance, heat resistance, and workability and is readily available. Although SUS 316 is not the optimal material for ATF cladding, we think it is persuasive to use it to confirm the possibility of the swaging process.

2. Experimental

2.1. Swaging Method

The inner tube comprised Zircaloy-4 (outer diameter: 9.57 mm, inner diameter: 8.30 mm, length: 1 m) and the outer tube comprised SUS 316 (outer diameter: 11.90 mm; inner diameter: 9.82 mm; length: 1 m). The surfaces of Zircaloy-4 and SUS 316 tubes were cleaned by wiping with ethanol. A double tube was prepared by inserting the Zircaloy-4 tube into the SUS 316 tube. The inner tube was completely filled with water-soluble KNO_3 filler powder, so that the tube would not experience distortion during compression in the direction perpendicular to the tube surface and also so that the stress could be uniformly applied in all directions. The PST was synthesized by applying a pressure of 4 t/cm^2 in the direction of the central axis of the double tube. Meanwhile, it was difficult to quantify the magnitude of the tensile stress in the transverse direction, which arose from the 4 t/cm^2 compressive stress applied in the longitudinal direction. Each time the tube was passed through the shaft jig, the speed of the tube was manually controlled, so that the outer diameter of the outer tube was reduced by 0.5 mm of the total. Figure 1 shows a schematic of the swaging method for the double tube and the stress type applied to the tube during the process. Figure 2 shows that the inner surface of the outer tube and the outer surface of the inner tube adhere closely to each other. The compressive stress applied during swaging and the tensile force applied in the longitudinal direction result in a strong adhesive force. In addition, the thickness, inner and outer diameters, and length of the final tube can be adjusted, depending on the number of repetitions. The KNO_3 left inside the PST was removed by immersing the PST in water at room temperature.

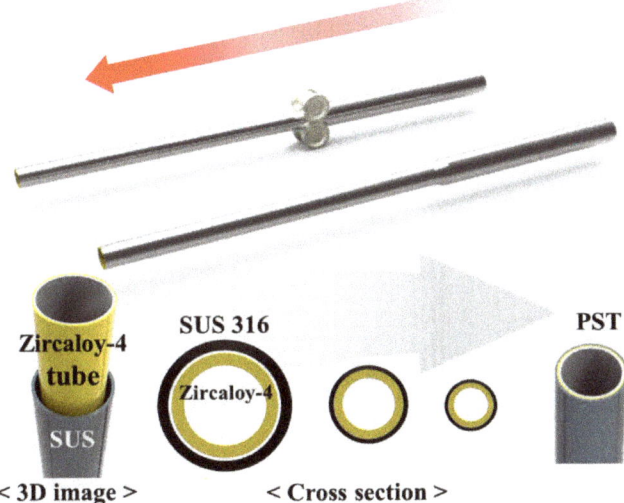

Figure 1. Schematic view of the swaging method for a double tube and the stress type applied to the tube during the process.

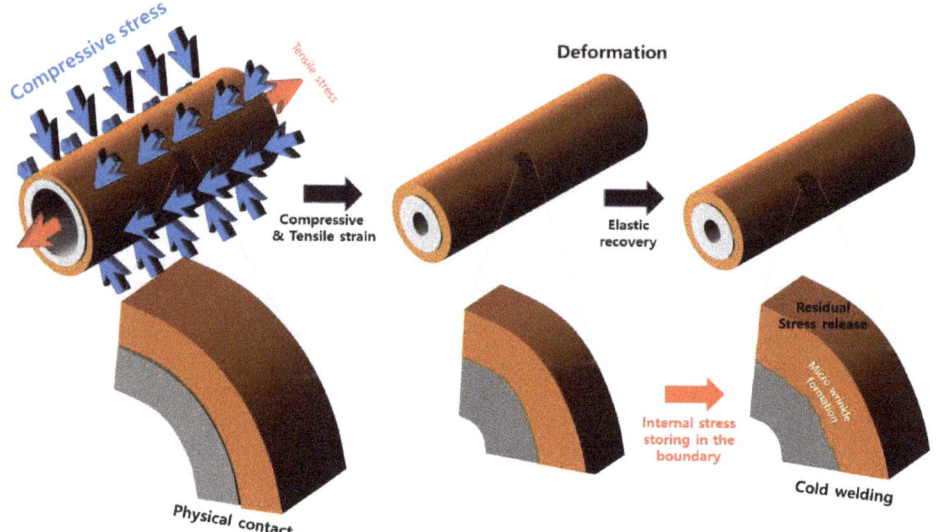

Figure 2. Mechanism of swaging.

2.2. Heat Treatment

Cut pieces of Zircaloy-4 cladding tubes and PSTs with a thickness of 2 cm were prepared and heat-treated in air under the following conditions: starting from room temperature (0~25 °C), after they had reached 600, 900, or 1200 °C, at a rate of 100 °C/h, each tube was maintained at the specified temperature for 1 h. The tubes were left to cool to 200 °C before being taken out of the furnace. Figure 3 shows a schematic of the heat treatment and the treatment conditions.

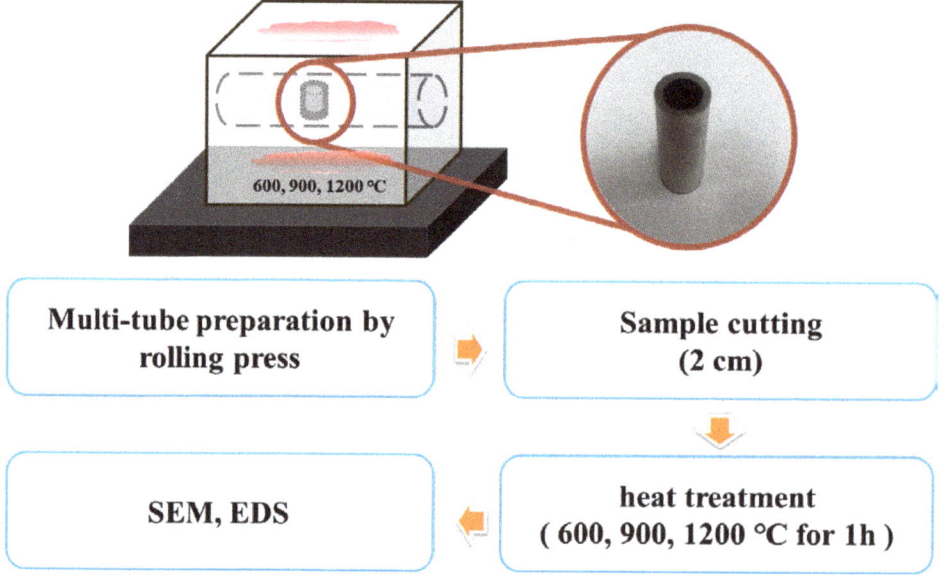

Figure 3. Schematic and conditions of heat treatment.

2.3. Analysis

To compare the degrees of oxidation and the stability of the interfaces after exposure to high temperatures, 2 cm-long tubes were prepared as follows: each tube was immersed in epoxy and dried until it solidified. Subsequently, the tubes immersed in the solidified epoxy were cut, their cross-sections were polished, and the microstructures of the polished surfaces were analyzed by OM. An SEM analysis was performed to examine the interfacial structure obtained from the OM at high resolution. EDS was used to analyze the diffusion behavior of the metal components at the PST interface after high-temperature heat treatment. In addition, because the two tubes were adhered solely by physical force, the cross-sections were analyzed using transmission electron microscopy (TEM) to confirm the characteristics of the interface at a higher resolution. The cross-sections of the PSTs were prepared for the TEM analysis by focused ion beam (FIB) milling.

3. Results and Discussion

Figure 4a shows the fabricated 90 cm-long PST and Figure 4b shows its cross-section before and after swaging. From Figure 4b it can be seen that some KNO_3 filler remained inside the tube with the smaller internal diameter after swaging, which could be easily removed with water. Figure 5 shows the OM image of three points on the interface of the PST. This image confirms that there were no gaps or voids in all three points of the interface between the two tubes, which was brought about by applying stress radially inward and lengthwise. Furthermore, the bonding interface between the two tubes was determined to be the same along the entire circumference. The uniformity of the interface of the PST was further confirmed by SEM images of the metal interface, as shown in Figure 6. It is imperative to note that gaps in the interface can degrade the mechanical and chemical properties of the tube. An interface without three-dimensional defects, such as gaps or voids, does not undergo oxidation reactions; therefore, it does not experience mechanical and chemical decomposition, even in high temperatures or corrosive environments [19]. As shown in Figure 6, the PST fabricated by swaging exhibits an exceedingly dense interface without three-dimensional defects.

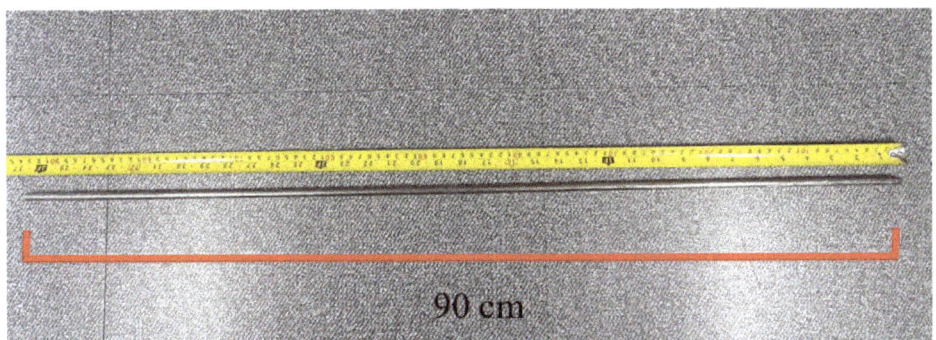

(a)

Figure 4. Cont.

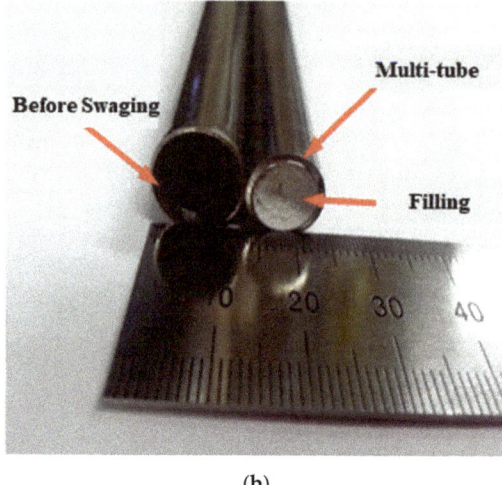

(b)

Figure 4. (**a**) Photographs of the final pseudo-single tube (PST), (**b**) Comparison of size change before and after swaging.

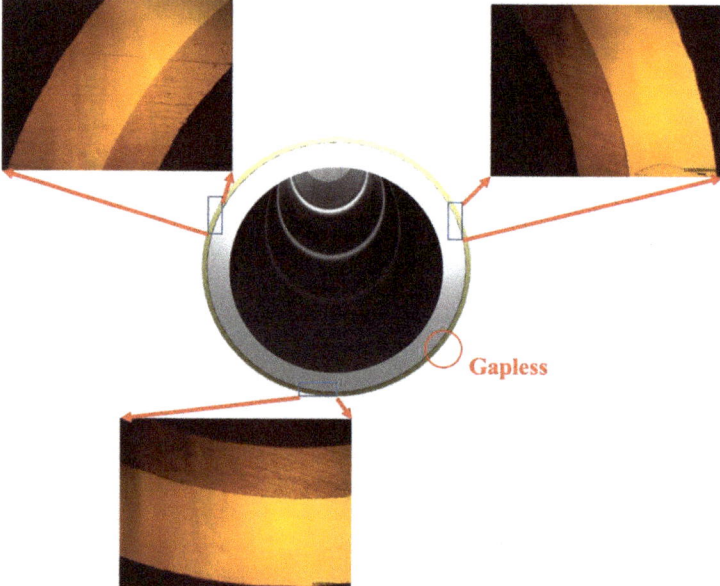

Figure 5. Optical microscopy image of PST without heat treatment. Schematic of the PST manufacturing process by swaging.

A peculiarity was that the tube thickness increased by approximately 5% from 3.35 to 3.52 mm after swaging, which can be explained as follows: during swaging, powder fillers were subjected to compressive stress in the direction of the center of the tube. Owing to plastic deformation, the inner diameter of the tube did not return to its original state, even after removal of the compressive stress [20]. During the initial stages of swaging, the tube length remained unchanged, thus increasing the tube thickness.

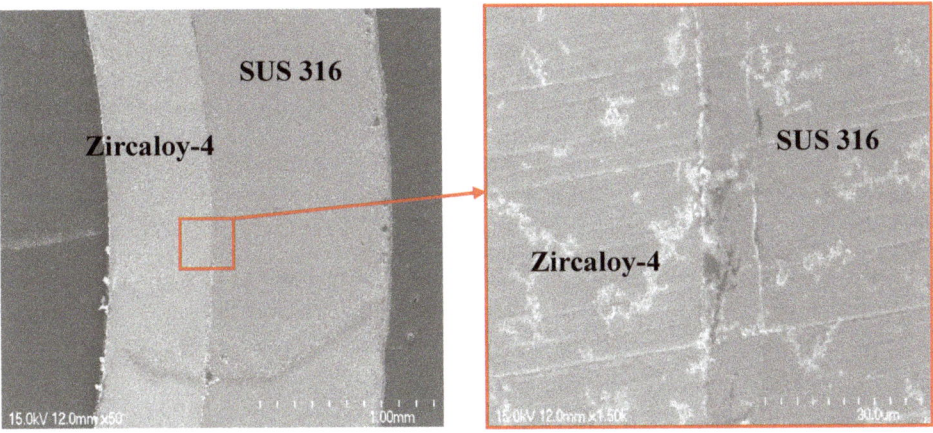

Figure 6. Secondary electron microscope image of the PST interface without heat treatment.

Once the filling powder entered a high-density state in which it could not be compressed further during swaging, the metal tube was also no longer compressed under external compressive stress. After this step, the tube started becoming longer and thinner due to the tensile stress applied to the tube. When using the swaging process, the magnitudes of the compressive stress acting perpendicular to the axial longitudinal direction and the tensile stress acting in the longitudinal direction were the most important variables for determining the length and thickness of the final PST. It is crucial to design a PST such that the initial sum of the thickness of each tube is as close as possible to the thickness of the single Zircaloy-4 tube currently in use. In particular, after completion of the process, the outer tube should ideally be suited to possess a thickness that modifies the chemical properties of the surface with little effect on the mechanical properties of the Zircaloy-4 and the degree of neutron absorption. Therefore, it is preferable to start the process with an outer tube as thin as possible. However, as mentioned above, in this study, the swaging process was carried out using a commercially available thick SUS 316 tube for the purpose of proving that Zircaloy-4 and dissimilar metals can be produced in the PST form using the described tube process. We are currently investigating the relationship between compressive stress and tube length and thickness.

At room temperature, the physical pressure applied toward the center of the tube and in the longitudinal direction of the two tubes leads to the formation of an interface with a strong bonding force. This can be explained by the mechanism, as illustrated in Figure 2. During swaging, the tube was subjected to compressive and tensile stresses in the axial and longitudinal directions. However, the tensile stress acting in the axial direction of the tube was not generated by the tube being pulled from the outside in the swaging equipment. This stress is thought to have been generated by the KNO_3 filler pressing toward the center. Tensile stress occurred in the longitudinal direction of the tube, as the inner diameter of the tube decreased, when a force of 4 t/cm^2 was applied to the center of the tube. This can be understood from the fact that the tube length increased after swaging, even if no tensile stress was applied. The most frequently mentioned cladding materials are Cr or Cr-based alloys and studies to optimize them are still incomplete. Because of this, it is not yet possible to measure or calculate the exact magnitude of the tensile stress generated in the longitudinal direction by the reduction of the inner diameter of the tube with the equipment used in this study. Clearly, however, the most important variable in the shaft process is the control of the compressive stress applied in the direction of the center of the shaft. We are designing and manufacturing new tube equipment that can accurately measure and control the magnitude of the tensile stresses created in the longitudinal direction induced by compressive stresses, along with the values of the compressive stresses.

Differences in physical properties, such as in the moduli of elasticity, shear moduli, and tensile strengths of the two metals after the tube process, result in different shrinkage behavior after stretching. Table 1 shows various physical properties of Zircaloy-4 and SUS 316. The total strain generated by swaging consists of a combination of elastic strain and plastic strain. An elastic deformation results in a return to the original state when the applied stress is removed; however, some permanent deformation, that is, plastic deformation, remained as the yield strength was exceeded. Finally, this plastic deformation resulted in a PST which is longer than the initial tube and has a smaller inner diameter than the initial tube [21]. As can be inferred from Table 1, Zircaloy-4 and SUS 316 tubes have different yield strengths, which can result in different final shrinkages. The first tube was 90 cm in length but post the tube process, the lengths of the Zircaloy-4 and SUS 316 tubes increased to 97.5 and 95.4 cm, respectively. Table 2 shows the sizes of the tubes before and after swaging. It can be seen that the outer and inner diameters decrease after the shaft.

Table 1. Physical properties of Zircaloy-4 and stainless use steel (SUS) 316.

Physical Properties	Zircaloy-4	SUS 316
Elastic modulus	99.3 GPa	193 GPa
Shear modulus	36.2 GPa	74–82 GPa
Tensile strength	514 MPa	580 MPa
Thermal expansion coefficient	6.3×10^{-6}/K	18.2×10^{-6}/K

Table 2. Outer diameters and inner diameters of the metal tubes before and after swaging.

Diameter	Zircaloy-4	SUS 316	PST
Outer diameter (mm)	9.57	11.90	11.60
Inner diameter (mm)	8.30	9.82	8.08

Thus, SUS 316, which experienced a relatively small plastic deformation, shrank more than Zircaloy-4. It is likely that the residual stress generated at the interface between the metal surface with a low plastic strain (high shrinkage) and the metal surface with a high plastic strain (low shrinkage) caused the two metal surfaces to adhere strongly to each other. In other words, SUS 316, which experienced a high degree of shrinkage, generated compressive stress on the surface of Zircaloy-4, which experienced only a low degree of shrinkage. Meanwhile, tensile stress was generated on the surface of SUS 316. The mechanism of the strong physical adhesion of the tubes, owing to the different residual stresses generated by the shaft tube process, is shown in Figure 7. This mechanism leads to a stable interfacial structure between the metal surfaces, as shown in the cross-sectional SEM image of the tube after swaging (Figure 6). Upon close observation of the interface between the two metals in Figure 6, it is evident that there is a small corrugated interfacial structure. In other words, the compressive stress generated on the surface of Zircaloy-4 due to the different shrinkage amounts of the two metals caused the Zircaloy-4 surface to shrink again, thereby forming wrinkles on the surface. This corrugated structure may have resulted in the grasping of the surface of SUS 316, which in turn led to the bonding of the two different metals after room temperature condensation. Such strong adhesive interface behavior was well demonstrated with quenching heat treatment experiments after high-temperature exposure, as shown below.

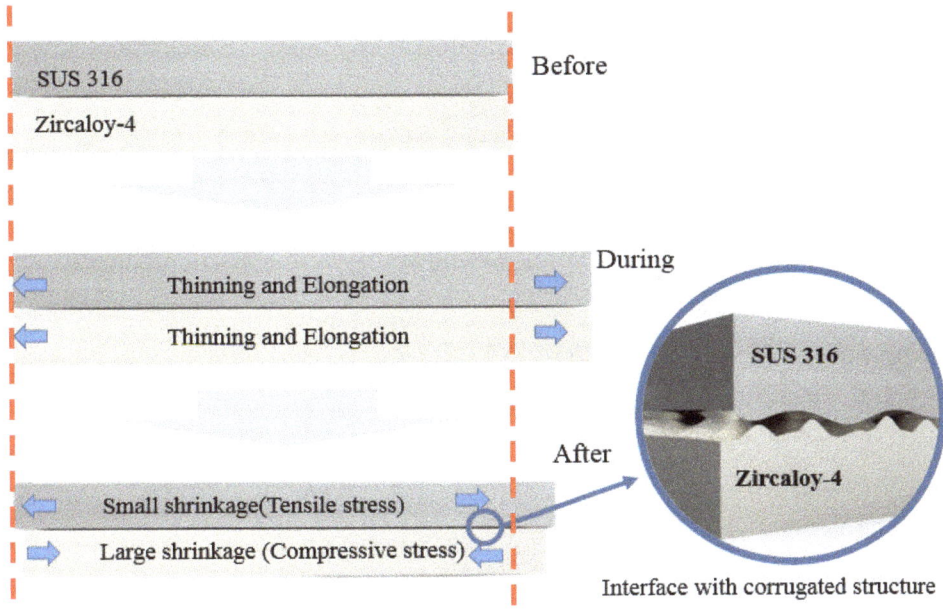

Figure 7. Mechanism for strong physical adhesion of different tubes due to residual stresses generated by the shaft tube process.

To evaluate the thermal stability of the PST fabricated using the room-temperature condensation process, an experiment was performed where the PST underwent rapid cooling after exposure to high temperature under ambient conditions. Figure 8 shows an image of a Zircaloy-4 tube and a PST quenched to room temperature after being maintained at temperatures of 600, 900, or 1200 °C for 1 h. The color of the bare Zircaloy-4 cladding changed due to oxidation, even at 600 °C. Under 1200 °C, the PST disintegrated not only due to severe oxidation phenomenon but also experienced external surface peeling [22].

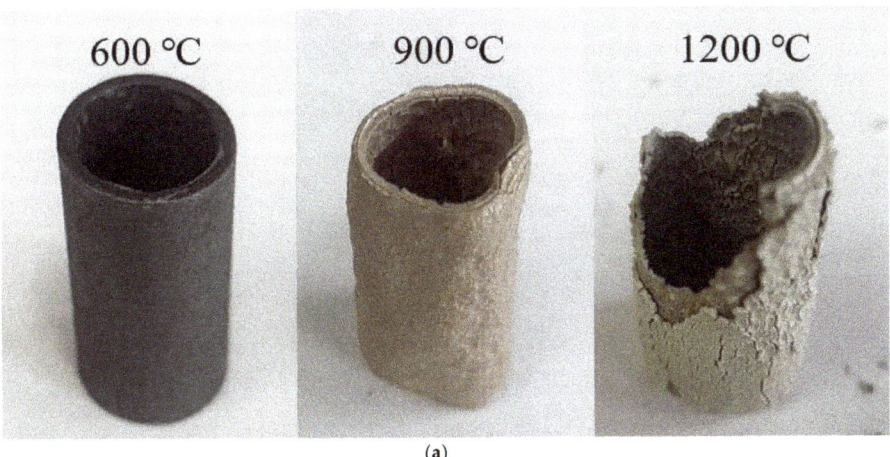

(a)

Figure 8. *Cont.*

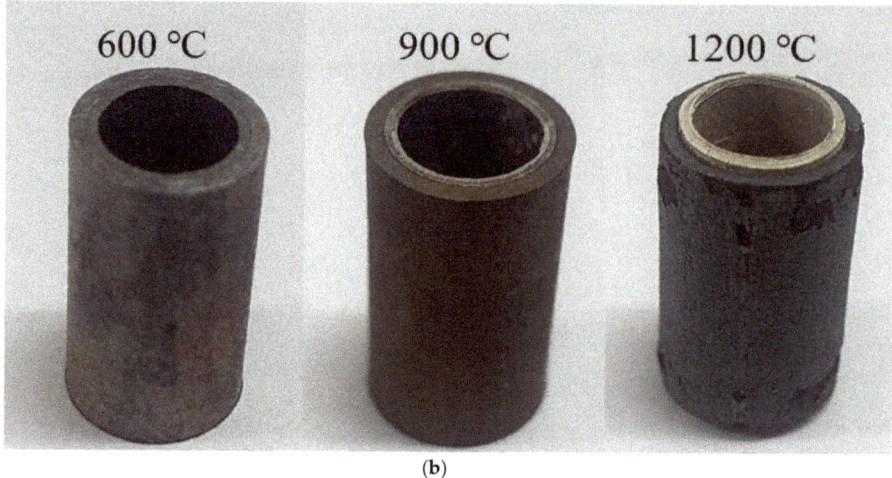

Figure 8. Appearance after maintaining (**a**) the Zircaloy-4-only tube and (**b**) the PST at 600, 900, and 1200 °C for 1 h.

Noticeably, the structural breakdown of the Zircaloy-4 tube after being exposed to high temperatures is exceedingly similar to the high-temperature disintegration of natural limestone [23]. Thermal shock tests of natural limestone concluded that its oxide layer is formed by an oxidation reaction at high temperatures. As the resulting oxide layer exhibits a large thermal expansion difference compared with the non-oxidized layer inside (owing to the differences in thermal shock received during cooling) and an immensely high amount of thermal stress accumulates at the interface between the oxide layer and the non-oxidized layer, the oxidized surface layer can be peeled off easily. In contrast, for the PST manufactured by the tube process, no surface peeling or structural failure was observed, even after heat treatment at 1200 °C. This is because the external material, SUS 316, which has a higher oxidation resistance than Zircaloy-4, inhibited the oxidation of Zircaloy-4 by preventing oxygen migration to the surface of Zircaloy-4. To observe the interface more clearly, the SEM analysis results of the PST cross-section before and after the heat treatment are shown in Figure 9. No collapse was observed at the interface of the quenched sample after exposure at 600 and 900 °C. In contrast, at 1200 °C, the interface of the quenched sample showed exfoliation and porosity due to volume expansion. It is assumed that this is caused by a large amount of oxygen diffusion from the center of Zircaloy-4 to the interface between the two metals. However, even when the PST was exposed to 1200 °C, the oxidation reaction of the Zircaloy-4 surface was suppressed, along with subsequent structural collapse.

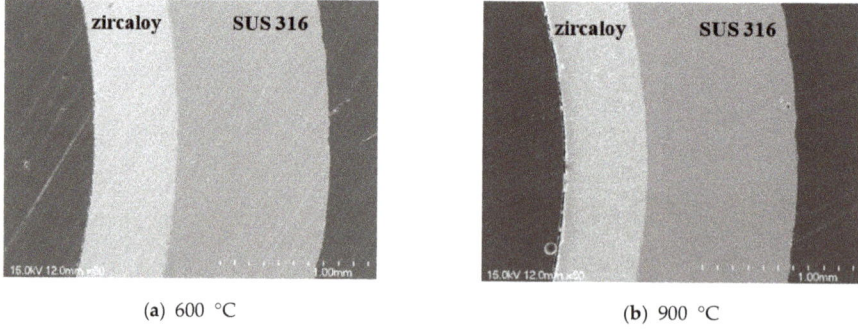

(**a**) 600 °C　　　　　　　　　　　　　　(**b**) 900 °C

Figure 9. *Cont.*

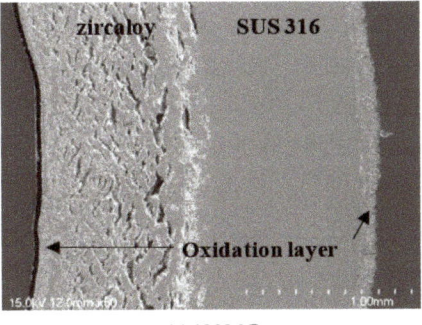

(c) 1200 °C

Figure 9. Scanning electron microscopy (SEM) images of cross-sections of the PST maintained at 600, 900, and 1200 °C for 1 h.

To confirm the diffusion behavior of the PST, an EDS analysis was conducted, as shown in Figure 10. After heat treatment at 600 and 900 °C, the diffusion of Zr did not occur in the SUS 316 direction and no metallic elements of SUS 316 were seen in the Zircaloy-4 direction. Compositional analysis by EDS line scanning showed that the interface between Zircaloy-4 and PST did not degrade even when exposed to high temperatures. However, after heat treatment at 1200 °C, a well-known eutectic reaction caused an interdiffusion reaction, in which an Fe-Zr complex with a melting point of 928 °C was formed [24]. Slight diffusion occurred at 1200 °C but it still exhibited high stability at operating and accident conditions. These results show that a PST consisting of a metal surface and Zircaloy-4 (e.g., 10Al-90Cr, etc.) [25], and fabricated using the tube process, has the potential to be used as ATF cladding. The interface of the sample, when exposed to 1200 °C, as well 600 °C and 900 °C, did not appear to peel off easily. From this, it can be concluded that an ATF cladding comprising the PST can maintain stability even at 1200 °C because the inside of the tube is isolated from external conditions during actual use. TEM analysis was performed to confirm the interfacial stability of the PST after exposure to high temperatures. Figure 11 is a cross-sectional TEM image that shows the interfacial structure of Zircaloy-4/SUS 316 of the PST when quenched to room temperature after being maintained at 900 °C for 1 h. Rapid cooling to room temperature after exposure to high temperatures is generally well-suited for inducing significant thermal stresses at an interface to observe the physical stability of the interface. Images were obtained at three different magnifications to observe the interfacial condition over a wide range of PST cross-sections and nanoscale. The width of the PST interface was 0.03 μm (30 nm) after heat treatment, as elucidated from the highest-magnification images. Moreover, no new interface was formed between the secondary phase and each tube due to the interdiffusion of the two tube components. This is in good agreement with the results of EDS, shown in Figure 10. Therefore, it is evident that the interface between two different tubes can maintain physicochemical stability, even when the PST formed by swaging is exposed to high temperatures. These results suggest that the development of several m-grade ATF claddings based on Zircaloy-4 is feasible, if conditions such as the type of tube used, its physical shape, and the optimum swaging pressure applied to the tube are established.

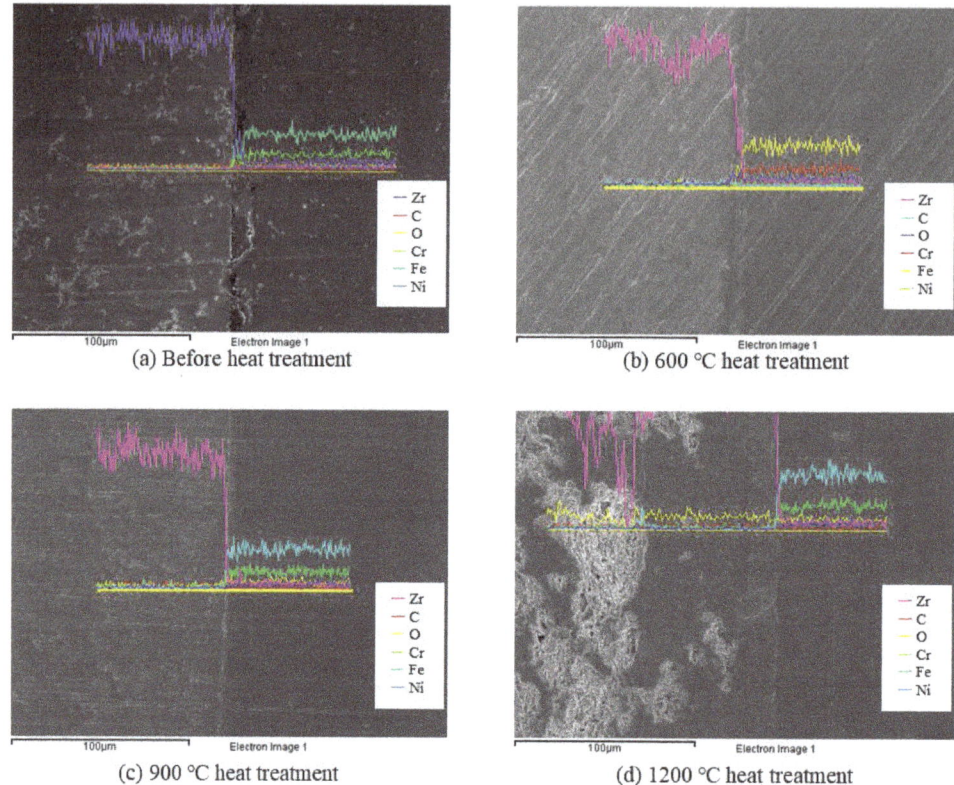

Figure 10. Energy-dispersive X-ray spectra of the PST before and after heat treatments at 600, 900, and 1200 °C for 1 h.

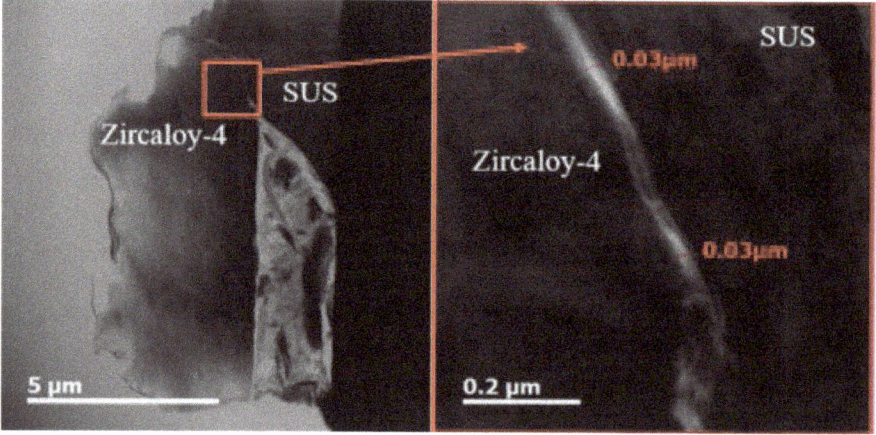

Figure 11. Cross-sectional transmission electron microscopy (TEM) image showing the interfacial structure of the Zircaloy-4/SUS 316 PST quenched to room temperature after being maintained at 900 °C for 1 h.

The results show that the tube process has a high potential for the development of an ATF cladding with a length of several meters, based on Zircaloy-4. The results suggest that controlling the compressive stress applied along the axial direction of the tube and the tensile stress occurring in the longitudinal direction will enable the facile and rapid development of ATF claddings, with their geometries calculated according to the design.

4. Conclusions

A Zircaloy-4-based PST, over 90 cm in length, was successfully fabricated using a simple straight tube process that was carried out at room temperature. Zircaloy-4, a material exhibiting well-known physicochemical properties for nuclear power operation, was inserted into a SUS 316 tube. It was possible to fix the SUS 316 tube to the outer surface of Zircaloy-4 by applying compressive stress toward the center of the tube, such that it exhibited strong adhesion. PST cross-sectional observations using OM and SEM revealed no three-dimensional defects, such as gaps or voids, at the interfaces of the two metal tubes. To confirm the high-temperature stability and thermal shock behavior of the PST, it was quenched to room temperature after being kept for 1 h at 600, 900, and 1200 °C. It was inferred from the OM and SEM analyses that no macroscopic defects were formed at the interface between SUS 316 and the Zircaloy-4 cladding after heat treatment. Even if the prepared PST was maintained at 900 °C, an interdiffusion reaction between the elements constituting Zircaloy-4 and SUS 316 would not occur. The Zircaloy-4-based PST fabricated using a room-temperature condensation process has been found to have highly stable physicochemical properties, even on quenching from high temperatures. Although SUS 316 is not the optimal cladding material, PST using this showed better thermal stability than bare Zircaloy-4 tube and it was confirmed that the swaging process can be applied to ATF cladding production. In conclusion, if (1) a thin external dissimilar metal tube and (2) a conduit device capable of controlling the compressive stress applied during the condensation process and the tensile stress in the longitudinal direction are used, then ATF claddings, which are widely used in nuclear power plant operations, can easily be manufactured to a length of up to 4 m. In addition, the swaging method described herein is likely to find potential use for a variety of applications, especially in fields where PST structures are crucial, i.e., when specific surface properties and uniform physical properties throughout the tube are critical.

Author Contributions: Conceptualization, Y.S.Y.; methodology, Y.S.Y. and J.H.K.; software, J.H.K.; validation, Y.K. and J.H.K.; formal analysis, D.Y.K. and Y.S.Y.; investigation, D.Y.K. and Y.N.L.; resources, Y.K.; data curation, J.H.K. and Y.K.; writing—original draft preparation, D.Y.K.; writing—review and editing, Y.N.L. and Y.S.Y.; visualization, Y.S.Y.; supervision, Y.K. and Y.S.Y.; project administration, Y.S.Y.; funding acquisition, Y.K. All authors have read and agreed to the published version of the manuscript.

Funding: This research was funded by Engineering Research Center of Excellence Program of Korea Ministry of Science, ICT and Future Planning (MSIP)/National Research Foundation of Korea (NRF) grant number NRF-2008-0062609 and was funded by Korea Hydro and Nuclear Power Co. Ltd. (KHNP) grant number No. 2018-safety-10.

Acknowledgments: This work was supported by the Engineering Research Center of Excellence Program of Korea Ministry of Science, ICT and Future Planning (MSIP)/National Research Foundation of Korea (NRF) (Grant NRF-2008-0062609). This work was also supported by Korea Hydro and Nuclear Power Co. Ltd. (No. 2018-safety-10).

Conflicts of Interest: The authors declare no conflict of interest.

References

1. Carmack, J.; Goldner, F.; Bragg-Sitton, S.M.; Snead, L.L. *Overview of the US DOE Accident Tolerant Fuel Development Program*; Idaho National Laboratory: Idaho Falls, ID, USA, 2013.
2. Cheng, B.; Kim, Y.J.; Chou, P.; Deshon, J. Development of Mo-alloy for LWR fuel cladding to enhance fuel tolerance to severe accidents. *Top Fuel* **2013**, *2013*, 15–19.

3. Idarraga-Trujillo, I.; Le Flem, M.; Brachet, J.C.; Le Saux, M.; Hamon, D.; Muller, S.; Vandenberghe, V.; Tupin, M.; Papin, E.; Monsifrot, E.; et al. Assessment at CEA of coated nuclear fuel cladding for LWRs with increased margins in LOCA and beyond LOCA conditions. *Top Fuel* **2013**, *2*, 15–19.
4. Kim, H.G.; Kim, I.H.; Park, J.Y.; Koo, Y.H. Application of Coating Technology on Zirconium-Based Alloy to Decrease High-Temperature Oxidation. *Zircon. Nucl. Ind.* **2015**, *465*, 346–369.
5. Bischoff, J.; Delafoy, C.; Vauglin, C.; Barberis, P.; Roubeyrie, C.; Perche, D.; Duthoo, D.; Schuster, F.; Brachet, J.C.; Schweitzer, E.W.; et al. AREVA NP's enhanced accident-tolerant fuel developments: Focus on Cr-coated M5 cladding. *Nucl. Eng. Technol.* **2018**, *50*, 223–228. [CrossRef]
6. Thompson, T.; Zachary, A.; Kurt, T.; Yukinori, Y. Elastic Modulus Measurement of ORNL ATF FeCrAl Alloys. *Ornl/Tm* **2015**, *632*, 1–17.
7. Wu, X.; Kozlowski, T.; Hales, J.D. Neutronics and fuel performance evaluation of accident tolerant FeCrAl cladding under normal operation conditions. *Ann. Nucl. Energy* **2015**, *85*, 763–775. [CrossRef]
8. Petrie, C.M.; Koyanagi, T.; McDuffe, J.L.; Deck, C.P.; Katoh, Y.; Terrani, K.A. Experimental design and analysis for irradiation of SiC/SiC composite tubes under a prototypic high heat flux. *J. Nucl. Mater.* **2017**, *491*, 94–104. [CrossRef]
9. Kim, H.G.; Kim, I.H.; Jung, Y.I.; Park, D.J.; Park, J.Y.; Koo, Y.H. Adhesion property and high-temperature oxidation behavior of Cr-coated Zircaloy-4 cladding tube prepared by 3D laser coating. *J. Nucl. Mater.* **2015**, *465*, 531–539. [CrossRef]
10. Bischoff, J.; Delafoy, C.; Chaari, N.; Vauglin, C.; Buchanan, K.; Barberis, P.; Monsifrot, E.; Schuster, F.; Brachet, J.C.; Nimishakavi, K. CR-coated cladding development at framatome. *Top Fuel* **2018**, *2018*, A0152.
11. Dryepondt, S.; Unocic, K.A.; Hoelzer, D.T.; Massey, C.P.; Pint, B.A. Development of low-Cr ODS FeCrAl alloys for accident-tolerant fuel cladding. *J. Nucl. Mater.* **2018**, *501*, 59–71. [CrossRef]
12. Rebak, R.B.; Gassmann, W.P.; Terrani, K.A. Managing nuclear power plant safety with FeCrAl alloy fuel cladding. In *Top Safe 2017*; IAEA Safety in Reactor Operation: Vienna, Austria, 2017.
13. Kim, D.; Ko, M.; Lee, H.G.; Park, J.Y.; Kim, W.J. Influence of Winding Patterns and Infiltration Parameters on Chemical Vapor Infiltration Behaviors of SiCf/SiC Composites. *J. Korean Ceram.* **2014**, *51*, 453. [CrossRef]
14. Kim, J.Y.; Oh, J.Y.; Lee, T.I. Multi-dimensional nanocomposites for stretchable thermoelectric applications. *Appl. Phys. Lett.* **2019**, *114*, 043902. [CrossRef]
15. Lee, J.S.; Shin, D.H.; Jang, J. Polypyrrole-coated manganese dioxide with multiscale architectures for ultrahigh capacity energy storage. *Energy Environ. Sci.* **2015**, *8*, 3030–3039. [CrossRef]
16. Kim, D.; Kim, J.; Lee, J.; Kang, M.K.; Kim, S.; Park, S.H.; Kim, J.; Choa, Y.H.; Lim, J.H. Enhanced Magnetic Properties of FeCo Alloys by Two-Step Electroless Plating. *J. Electrochem. Soc.* **2019**, *166*, D131–D136. [CrossRef]
17. Ferrer-Argemi, L.; Yu, Z.; Kim, J.; Myung, N.V.; Lim, J.H.; Lee, J. Silver content dependent thermal conductivity and thermoelectric properties of electrodeposited antimony telluride thin films. *Sci. Rep.* **2019**, *9*, 1–8. [CrossRef]
18. Nam, H.; Kim, H.S.; Han, J.H.; Kwon, S.J.; Cho, E.S. A Study on the Formation of 2-Dimensional Tungsten Disulfide Thin Films on Sapphire Substrate by Sputtering and High Temperature Rapid Thermal Processing. *J. Nanosci. Nanotechnol.* **2018**, *18*, 6257–6264. [CrossRef] [PubMed]
19. Giggins, C.S.; Pettit, F.S. Corrosion of metals and alloys in mixed gas environments at elevated temperatures. *Oxid. Met.* **1980**, *14*, 363–413. [CrossRef]
20. Hofmann, P.; Markiewicz, M. Chemical interactions between as-received and pre-oxidized Zircaloy-4 and stainless steel at high temperatures, No. KFK–5106. Kernforschungszentrum Karlsruhe GmbH (Germany). *Proj. Nukl. Sicherh.* **1994**, *26*, 1–41.
21. Wu, H.Y.; Lee, S.; Wang, J.Y. Solid-state bonding of iron-based alloys, steel–brass, and aluminum alloys. *J. Mater. Process. Technol.* **1998**, *75*, 173–179. [CrossRef]
22. Kim, H.H.; Kim, J.H.; Moon, J.Y.; Lee, H.S.; Kim, J.J.; Chai, Y.S. High-temperature oxidation behavior of Zircaloy-4 and Zirlo in steam ambient. *J. Mater. Sci. Technol.* **2010**, *26*, 827–832. [CrossRef]
23. Becattini, V.; Motmans, T.; Zappone, A.; Madonna, C.; Haselbacher, A.; Steinfeld, A. Experimental investigation of the thermal and mechanical stability of rocks for high-temperature thermal-energy storage. *Appl. Energy* **2017**, *203*, 373–389. [CrossRef]

24. Kim, J.H.; Choi, B.K.; Baek, J.H.; Jeong, Y.H. Effects of oxide and hydrogen on the behavior of Zircaloy-4 cladding during the loss of the coolant accident (LOCA). *Nucl. Eng. Des.* **2006**, *236*, 2386–2393. [CrossRef]
25. Kim, J.M.; Ha, T.H.; Kim, I.H.; Kim, H.G. Microstructure and oxidation behavior of CrAl laser-coated Zircaloy-4 alloy. *Metals* **2017**, *7*, 59. [CrossRef]

© 2020 by the authors. Licensee MDPI, Basel, Switzerland. This article is an open access article distributed under the terms and conditions of the Creative Commons Attribution (CC BY) license (http://creativecommons.org/licenses/by/4.0/).

Article

Application of the 2-D Trefftz Method for Identification of Flow Boiling Heat Transfer Coefficient in a Rectangular MiniChannel

Mirosław Grabowski [1],*, Sylwia Hożejowska [2], Beata Maciejewska [2], Krzysztof Płaczkowski [1] and Mieczysław E. Poniewski [1]

[1] Faculty of Civil Engineering, Mechanics and Petrochemistry, Warsaw University of Technology, 09-400 Płock, Poland; liloslaw@wp.pl (K.P.); mieczyslaw.poniewski@pw.edu.pl (M.E.P.)
[2] Faculty of Management and Computer Modelling, Kielce University of Technology, 25-314 Kielce, Poland; ztpsf@tu.kielce.pl (S.H.); beatam@tu.kielce.pl (B.M.)
* Correspondence: miroslaw.grabowski@pw.edu.pl

Received: 25 May 2020; Accepted: 21 July 2020; Published: 2 August 2020

Abstract: The study presents the experimental and numeric heat transfer investigations in flow boiling of water through an asymmetrically heated, rectangular and horizontal minichannel, with transparent side walls. A dedicated system was designed to record images of two-phase flow structures using a high-speed video camera with a synchronous movement system. The images were analyzed with Matlab 2019a scripts for determination of the void fraction for each pattern of two-phase flow structures observed. The experimental data measured during the experimental runs included inlet and outlet temperature, temperature at three internal points of the heater body, volume flux of the flowing water, inlet pressure, pressure drop, current and the voltage drop in the heater power supply. The flows were investigated at Reynolds number characteristic of laminar flow. The mathematical model assumed the heat transfer process in the measurement module to be steady-state with temperature independent thermal properties of solids and flowing fluid. The defined two inverse heat transfer problems were solved with the Trefftz method with two sets of T- functions. Graphs were used to represent: the boiling curves, the local void fraction values, the boiling heat transfer coefficients and the errors of both of them for selected mass fluxes and heat fluxes.

Keywords: minichannel flow boiling; void fraction; inverse heat transfer problem; Trefftz method

1. Introduction

A growing necessity of transferring very high heat fluxes from both miniaturized industrial equipment and home appliances generates demand for mini- or micro-scale cooling devices. Some cooling systems require low pumping power and consequently low Reynolds numbers for the flow [1–3]. In such cases, flow boiling heat transfer, which is characteristic of high heat transfer coefficients, appears to be the appropriate solution. The determination of the heat transfer coefficient requires knowledge of the parameters of the boiling fluid flowing in the minichannel. These parameters are in particular: temperature and pressure at characteristic points of the system, mass flux of the boiling fluid, heat flux and void fraction. The measurements of the void fraction were combined with photographic recording of boiling two-phase flow structures, characteristic for horizontal orientation of the minichannel [4,5].

The obtained experimental data for water flow boiling in a single rectangular minichannel at a low Reynolds number provided the basis for numeric analysis with the use of Trefftz method [6]. Direct heat conduction problems—DHCPs—Appear in mathematical modeling of steady-state heat transfer problems when the governing equation, domain, boundary conditions and physical properties of the material are known. When any of these elements is unknown, we must deal with an inverse heat conduction problem—IHCP. High sensitivity to uncertainty of incoming data are a characteristic feature of many engineering inverse problems. This weakness can get enhanced when two or three succeeding inverse problems are taken into account. Therefore, solving inverse problems requires efficient and stable numeric methods. As the Trefftz method with a set of T-functions fulfills this requirement [7–11], it was used to solve the IHCP problem for flow boiling in the minichannel. Two sets of T-functions were applied to calculate: (a) two-dimensional temperature distributions of the heating copper block and the boiling water, (b) heating copper block temperature gradients, and (c) the heat transfer coefficient at the contact surface copper block–flowing fluid.

2. Experimental Facility

2.1. Design of the Flow Loop, Experimental Equipment and Data Collecting Procedure

The flow loop and its components are presented in Figure 1.

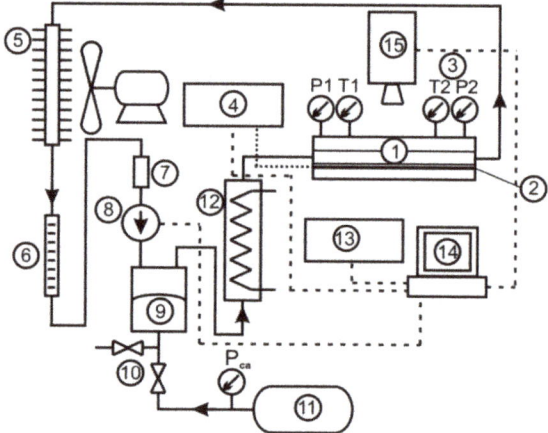

Figure 1. Flow loop. 1—Measurement module with the minichannel (details in Figure 3), 2—Heating copper block, 3—Temperature and pressure sensors, 4 DC power supply, 5—Cooler with ventilator, 6—Rotameter, 7—Filter, 8—Pump, 9—Pressure control, 10—Compressed air valves, 11—Compressed air tank, 12—Preheater, 13—Facility control unit, 14—Computer for experiment control + LabView software, 15—High speed camera, P_{ca}—Compressed air pressure sensor.

The minichannel, the basic element of the experimental stand (Figure 2), was constructed by glue bonding three transparent glass plates and a cuboid copper block, Figure 3. The cuboid copper block was a solid base for the minichannel and a mass heater as well. Four flat resistance heaters were placed on steel substrates and sintered to the external surface of the copper block. A TDK Lambda GEN 50-30 power supply provided direct current to the heaters, which generated heat required to initiate flow boiling inside the channel (Figures 1 and 2).

Optiwhite glass was selected for the minichannel walls. Optiwhite is colorless, super transparent float glass containing very limited amount of iron and having the highest light-transmission coefficients. Three transparent walls of the channel enabled proper lighting and recording the boiling two-phase structures with a high-speed camera (Phantom 711, Vision Research).

To prevent uncontrolled heating of the flowing fluid by the incandescent light, the LED system of our own design, based on LED diode Citizen CL-L233-HC13L1-C, was applied. LOCTITE® SI 5145 adhesive was used to glue the minichannel elements. The dimensions of the rectangular minichannel, length = 180 mm, width = 4 mm and depth = 1.5 mm, provided a 6 mm² cross-section and a 2.18 mm hydraulic diameter. Five thermocouples (Czaki TP-201) were mounted, one at the inlet, one at the outlet of the minichannel and three inside the body of the cuboid copper block, Figure 3. Two pressure sensors (Kobold 0–2.5 bar) were placed at the inlet and outlet of the channel. The flow of distilled water was generated by a precision gear pump (Tuthill Concord D) with a maximum capacity of 1.5×10^{-7} m^3/s, featuring very stable laminar flow in the range $43 \leq Re \leq 229$. Flows at low Reynolds number are quite frequently used in miniature cooling systems of electronic devices.

Figure 2. General view of the experimental stand; labels as in Figure 1.

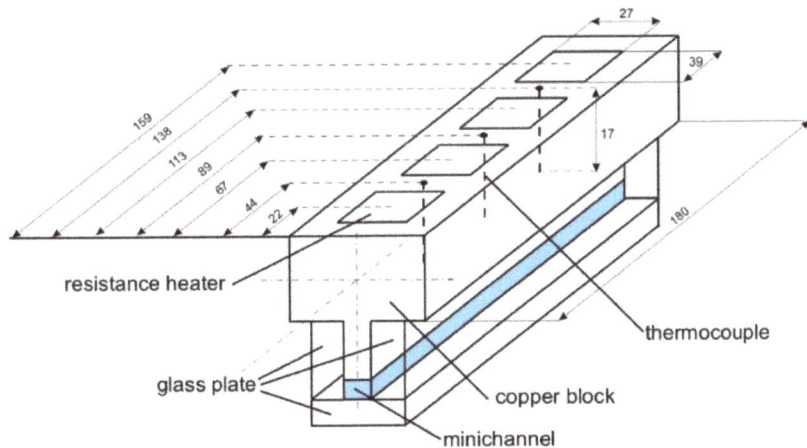

Figure 3. Copper block with attached glass minichannel and location scheme of resistance heaters and thermocouples—Transverse section and axonometric view; dimensions in mm (pictorial view, not to scale).

In the experimental runs, it was necessary to record large numbers of data of various types. Those included video camera images of two-phase flow structures and numeric data of copper block and flowing fluid temperature, fluid pressure and volume flux. Data collection was done by the modular measurement-control system, Figure 4. The core of it was NI cDAQ-9178 chassis (National Instruments) that holds dependent modules, controlling both the measurements of boiling process parameters and the component assemblies, such as the pump, camera triggering, etc.

The following modules comprised the system, as shown in Figure 4: NI cDAQ-9178—the main module, NI 9214—Temperature measurement (CZAKI K-type TP-201 thermocouples), NI 9239—Voltage measurement (KOBOLD pressure sensors, 0–2.5 bar, NI 9203—Current measurement (KOBOLD pressure difference sensors, 0–2.5 bar), NI 9263—Voltage setting for pump control.

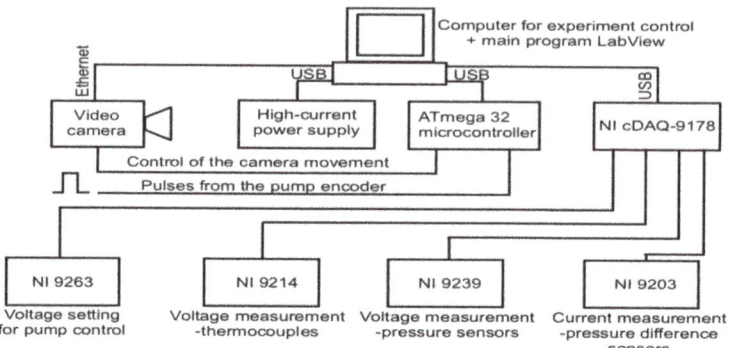

Figure 4. Block diagram of the data flow and control system.

In the supplementary measurements of heat loss to the environment, the entire minichannel module was treated as a cuboid with walls of different temperatures measured with the infrared camera. The calculations also accounted for thermal losses from the sight glass. The maximum value of the amount of heat released into the environment in the range of heat fluxes generated in the presented measurements did not exceed 2.1 W, which in relation to the total delivered thermal power was 0.86% to 1.9%. Since the heat losses were minor, they were disregarded in further calculations.

2.2. Procedure of Void Fraction Measurement and Computation in Rectangular Minichannel

Photographs of the observed two-phase flow structures were taken with a high-speed video camera at the recording speed of 7000 fps. The recording speed was selected experimentally to maintain proper exposure and obtain sharp images of dynamic two-phase flow structures. Proper selection of the recording speed is essential for capturing geometric features of the vapor bubbles. The equations approximating spatial geometry of vapor bubbles were applied to convert flat images of bubbles into three dimensional ones, taking simultaneously into account the dimensions of both the bubbles and the minichannel. The shapes of sphere and ellipsoidal cylinder ended with one or two half-ellipsoids were used, depending on inter relations between the dimensions of the bubbles and those of the minichannel. Three cases I, II and III of the characteristic relation between the bubble and minichannel dimensions were specified for void fraction determination in the observed minichannel of length L, width a and depth b, Figures 5–7.

Figure 5. (**A**) Actual and (**B**) approximated shapes of vapor bubbles in longitudinal and transverse sections of the channel with flow direction, case I. Water, $T_{in} = 63.1\ °C$, $T_{out} = 123.8\ °C$, $p_{in} = 1.165$ bar, $\Delta p = 0.0024$ bar, $q = 298.7$ kW/m^2, $G = 7.2$ kg/(m^2 s), $Re = 55.0$, $L_{cam} = 20$ mm.

Figure 6. (**A**) Actual and (**B**) approximated shapes of vapor bubbles in longitudinal and transverse sections of the channel with flow direction, case II. Water, $T_{in} = 55.2\ °C$, $T_{out} = 125.9\ °C$, $p_{in} = 1.0697$ bar, $\Delta p = 0.0016$ bar, $q = 260.1$ kW/m^2, $G = 8.6$ kg/(m^2 s), $Re = 66.9$, $L_{cam} = 20$ mm

Figure 7. (**A**) Actual and (**B**) approximated shapes of vapor bubbles in longitudinal and transverse sections of the channel with flow direction, case III. Water, T_{in} = 72 °C, T_{out} = 120 °C, p_{in} = 1.147 bar, Δp = 0.0016 bar, q = 270.5 kW/m², G = 5.9 kg/(m² s), Re = 43.5, L_{cam} = 20 mm.

Volume of the observed part of the minichannel was:

$$V = L \cdot a \cdot b, \tag{1}$$

1. **Small vapor bubbles: Rsb, i ≤ b/2.**

The void fraction for small bubbles in the minichannel was calculated from the equation [12]:

$$\varphi_{sb} = \frac{4}{3Lab\sqrt{\pi}} \sum_i A_{sb,i}^{\frac{3}{2}}, \tag{2}$$

where $A_{sb,i} = \pi R_{sb,i}^2$ is the cross sectional area of single spherical bubble.

2. Large, elongated bubbles, fully visible: ellipse semi-axes P1 $_{lb,\,i}$= $a/2$ and P2 $_{lb,\,i}$= $b/2$.

The void fraction for large, elongated and fully visible vapor bubbles was [12]:

$$\varphi_{lb} = \sum_i \varphi_{lb,i}, \tag{3}$$

where $\varphi_{lb,i} = \frac{(L_{lb,i}-a)\frac{\pi ab}{4} + \frac{\pi a^2 b}{6}}{Lab}$ is the void fraction for a single elongated bubble, $L_{lb,\,i}$ is the bubble length, composed of the length of an ellipsoidal cylinder and two half-ellipsoids.

3. Large, elongated bubbles, partially visible: ellipse semi-axes P1 $_{lb,\,i}$= $a/2$ and P2 $_{lb,\,i}$= $b/2$.

The void fraction for large, elongated and partially visible vapor bubbles was [12]:

$$\varphi_{lb} = \sum_i \varphi_{lb,i}, \tag{4}$$

where $\varphi_{lb,i} = \frac{\left(L_{lb,i}-\frac{a}{2}\right)\frac{\pi ab}{4} + \frac{\pi a^2 b}{12}}{Lab}$ is the void fraction for a single elongated bubble, $L_{lb,\,i}$ is the bubble length, composed of the length of an ellipsoidal cylinder and one half-ellipsoid.

Three scripts were developed to analyze the void fraction for selected cases of two-phase boiling flow structures found in the recorded videos. MathWorks Matlab and two corresponding toolboxes, computer system vision and image processing, were used for that purpose. By applying the first plan method and the Gaussian model, cases I and II were detected and elaborated with the use of the Vision toolbox). The background subtraction method appeared suitable for case III. Sharpening and pixel multiplication applied for each video frame improved quality of the recorded images. These operations were finally followed by equalization of brightness and contrast of the images to obtain a larger number of details, as shown in Figure 8.

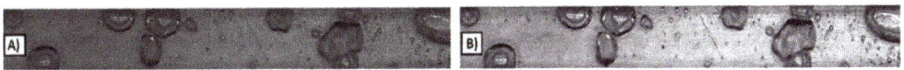

Figure 8. Picture of two-phase flow (**A**) before and (**B**) after the operations improving its quality.

Subsequently the perfected and binarized void frames underwent the following operations: morphologic opening and closing the picture, median filtration from artefacts and morphologic filling of empty spaces in the observed objects—That is vapor bubbles. The application of the entire algorithm converted the image in Figure 8B into the black and white one, Figure 9.

Figure 9. Image of two-phase flow resulting from morphologic operations.

Indexing and measuring procedure was used to get geometric dimensions of each detected bubble. On the basis of these data and with the use of equations presented in this section, by converting the flat picture of the bubble into the expected spatial one, the sought void fraction was calculated in the observed part of the minichannel. The last action of the script was recording the obtained void fraction value for each video frame into a text file for further use.

The estimated values of the void fraction were compared to seven correlations to calculate the void fraction against vapor quality. The results were similar [12]. The maximum scatter did not exceed 16%.

3. Mathematical Model and Numeric Solution

The mathematical model is a modification of the models described earlier in [2,13]. Analysis of the numeric calculation results presented in [13] showed that the change of the temperature along the width of the copper block body was negligible. In the two-dimensional model presented below, the copper block temperature depended only on two variables: y (along its height) and x (in relation to its length). Additionally, this assumption allowed considering heat conduction only in two elements of the measuring module: the copper block and the minichannel with flowing water.

By analogy to [13], we assumed that the heat transfer process in the experimental module was steady-state with constant copper and fluid properties. We also assumed that the temperature of the copper block in domain $D_1 = \{(x,y) \in R^2 : 0 < x < L, 0 < y < H_2\}$ satisfied the Laplace equation, i.e.,:

$$\nabla^2 T_C = 0. \tag{5}$$

For Equation (5), we assumed that the temperature of the copper block T_C was known at three measuring points (x_i, H_1), Figure 10.

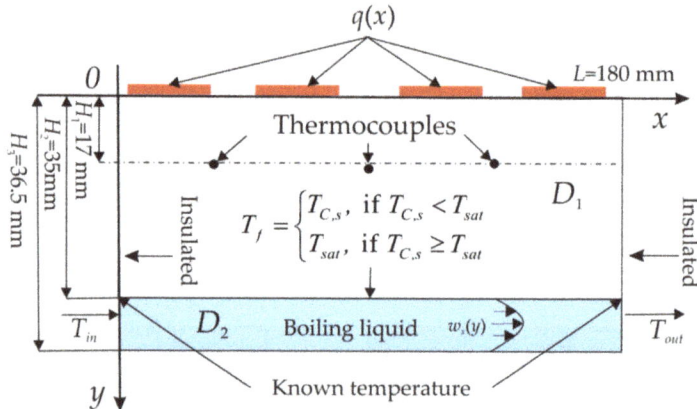

Figure 10. Scheme of the experimental stand with the adopted boundary conditions; (pictorial view, not to scale).

As in [13], in order to stabilize the numeric calculations two temperature values were added to boundary conditions at both ends of the fluid-copper heater contact surface:

a) $T_C(0, H_2) = 0.5(T_{in} + T_{approx}(0))$ b) $T_C(L, H_2) = 0.5(T_{out} + T_{approx}(L))$, (6)

where the broken line approximation of three measured temperatures inside the block, Figure 10, was used to obtain T_{approx}. Both vertical walls of the copper block were insulated.

The heat flux q generated by each resistance heater was supplied to the copper block, i.e.,:

$$\lambda_C \frac{\partial T_C(x,0)}{\partial y} = -q(x). \tag{7}$$

The heat flux $q(x)$ assumed constant values q_i, generated by a single heater in the segments, where four resistance heaters were located, Figure 10. Between these segments the heat flux equaled zero:

$$q(x) = \begin{cases} q_i & \text{for } x \in D_i, y = 0 \\ 0 & \text{for } x \in [0, L] - D, y = 0 \end{cases}, \tag{8}$$

where $D = \bigcup_{i=1}^{4} D_i = [2.5; 41.5] \cup [47.5; 86.5] \cup [93.5; 132.5] \cup [139.5; 178.5]$.

For fluid we assumed that, as shown in Figure 10:

- The flow in the horizontal minichannel was laminar ($Re < 2000$) and stationary with a constant volumetric flow rate,
- Liquid flow in the minichannel was a nonslip flow and the velocity of the fluid had only one non-zero parabolic component $w_x(y)$, parallel to the heating block and satisfying the condition:

$$\frac{1}{b} \int_{H_2}^{H_2+b} w_x(y) dy = w_{ave}, \tag{9}$$

where the w_{ave} was known
- The fluid temperatures at the inlet $T_{f,in}$ and outlet $T_{f,out}$ of the minichannel were known,
- The fluid temperature at the contact with the heater block fulfilled the condition:

$$T_f = \begin{cases} T_{C,s}, & \text{if } T_{C,s} < T_{sat} \\ T_{sat}, & \text{if } T_{C,s} \geq T_{sat} \end{cases}. \tag{10}$$

In domain $D_2 = \{(x, y) \in R^2 : 0 < x < L, H_2 < y < H_3\}$, Figure 10, the fluid temperature satisfied the energy equation:

$$\lambda_f \nabla^2 T_f = w_x(y) c_p \rho_f \frac{\partial T_f}{\partial x} - \mu_f \Phi - w_x(y) \frac{dp}{dx} + \Omega(x), \tag{11}$$

where function $\Phi = \left(\frac{d\, w_x}{dy}\right)^2$ was the Rayleigh dissipation function, pressure gradient $\frac{dp}{dx} \approx \frac{p_{out} - p_{in}}{L} = \frac{\Delta p}{L}$ and $\Omega(x)$ was negative heat source. The heat flux absorbed by the bubbles, also called a negative heat source, was calculated from the formula, which used experimentally determined void fraction $\phi(x)$ [14–16]:

$$\Omega(x) = \frac{6 \alpha_{con} \Delta T}{2R_b} \varphi(x). \tag{12}$$

The bubble diameter $2R_b$ in subcooled flow boiling was calculated using the correlation proposed in [17]:

$$2R_b = \min\left(1.4, 0.6 \cdot \exp\left(-\frac{\Delta T}{45}\right)\right) \cdot 10^{-3}. \tag{13}$$

The convective heat transfer coefficient α_{con} in Equation (12) was given by Labuncov correlation [18] for laminar flow:

$$\alpha_{con} = \frac{\lambda_f}{2R_b} 0.125 \text{Re}^{0.65} \text{Pr}^{\frac{1}{3}} \qquad (14)$$

and ΔT was the difference between the fluid temperature T_f in the thermal sublayer δ_T and the temperature inside the vapor bubble, which assumed to equal T_{sat}. Consequently, the fluid superheat ΔT was calculated from the dependence:

$$\Delta T = T_f - T_{sat} = \frac{q \cdot \delta_T}{\lambda_f}. \qquad (15)$$

In (15) the thermal and hydraulic layer thicknesses were calculated from formulas [16], respectively:

$$\Delta T = T_f - T_{sat} = \frac{q \cdot \delta_T}{\lambda_f}, \quad \delta_T = \text{Pr}^{-\frac{1}{3}} \delta_h, \quad \delta_h = \frac{2\mu}{f w_b \rho_f}, \qquad (16)$$

where $w_b(y) = w_x(y)$ and the Fanning friction factor was calculated as in [19]:

$$f\text{Re}_f = 24\left(1 - 1.3553\frac{b}{L} + 1.9467\left(\frac{b}{L}\right)^2 - 1.7012\left(\frac{b}{L}\right)^3 + 0.9564\left(\frac{b}{L}\right)^4 - 0.2537\left(\frac{b}{L}\right)^5\right). \qquad (17)$$

The known copper block temperature distributions and the temperature gradient were applied to determine the heat transfer coefficient $\alpha(x)$ at the copper–fluid interface using the Robin boundary condition:

$$\alpha(x) = \frac{-\lambda_C \frac{\partial T_C}{\partial y}(x, H_2)}{T_C(x, H_2) - T_{f,ave}(x)}. \qquad (18)$$

The reference temperature $T_{f,\,ave}$ was the average fluid temperature along the depth of the fluid layer equal to the diameter of emerging vapor bubble:

$$T_{f,ave}(x) = \frac{1}{2R_b} \int_{H_2}^{H_2+2R_b} T_f(x,y) dy. \qquad (19)$$

The energy Equations (5) and (11) with a set of the adopted boundary conditions led to the solution of two succeeding inverse heat conduction problems (IHCPs) in two domains of different shapes and parameters. The Trefftz method with two sets of adequate T-functions was used to solve the problem.

The boundary conditions Equations (6)–(8) and the harmonic functions (T-functions), defined in [6], were employed to solve Equation (5). Linear combination of T-functions approximated the copper block temperature. These coefficients were found by minimization of the mean square error between the approximated temperature and the boundary conditions. T-functions for homogeneous energy equation, corresponding to energy Equation (11) were applied to find the flowing fluid temperature [20].

The fluid temperature was approximated by a sum of particular solution of Equation (11) and the linear combination of the T-functions shown in [9]. Similar to the case of copper block temperature, the boundary conditions for the fluid were adopted to compute the coefficients of that linear combination.

4. Results and Discussion

The measurement procedure included recording the following quantities: temperature and pressure at the channel inlet and outlet, differences in inlet and outlet pressures, volume flux of the flowing medium, temperature at three selected points inside the copper block and also current and voltage of the heater power supply. The measurements were taken for the following ranges of the experimental parameters: total heat flux generated by four external flat heaters $188 \leq q \leq 340$ kW/m^2, inlet pressure $1.05 \leq p \leq 1.17$ bar (practically constant), inlet fluid subcooling $3.6 \leq \Delta T \leq 70.7$ K and mass flux $2.2 \leq G \leq 8.6$ kg/(m^2 s), $43 \leq Re \leq 229$.

At the same time, flow boiling in the minichannel was filmed using high-speed video camera and local void fractions $\varphi(x)$ were measured on the basis of recorded images. The scripts developed in Section 2.2 allowed the calculation of the void fraction for each pattern of two-phase flow structures observed in this study. The experimental local values of the void fraction, measured along the minichannel length, were approximated with broken lines, as shown in Figure 11.

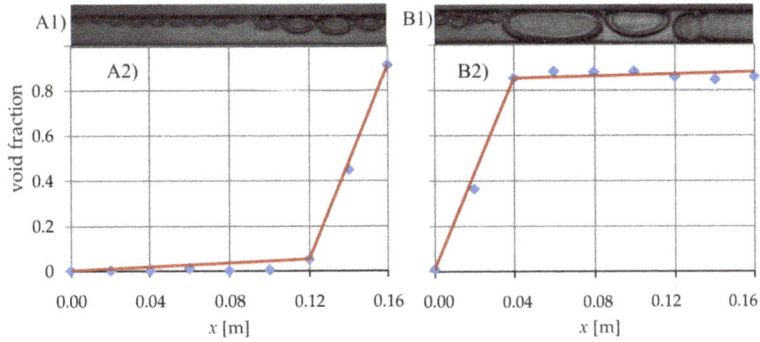

Figure 11. Examples of recorded flow structures presented in (**A1**): $q = 298.8$ kW/m^2, $G = 8.6$ kg/(m^2 s); (**B1**): $q = 223.2$ kW/m^2, $G = 2.2$ kg/(m^2 s). Corresponding void fractions (blue points), approximated with broken lines (red lines), are shown in (**A2**) and (**B2**).

The results of numeric calculations presented in Figure 12 were obtained for experimental parameters given in Section 4. In the first step, nine T-complete functions were used to calculate the approximate two-dimensional temperature of the copper block T_C for Equation (5). Next, by combining four T-complete functions for Equation (20) and particular solution of Equation (11), the field of the fluid temperature T_f was found.

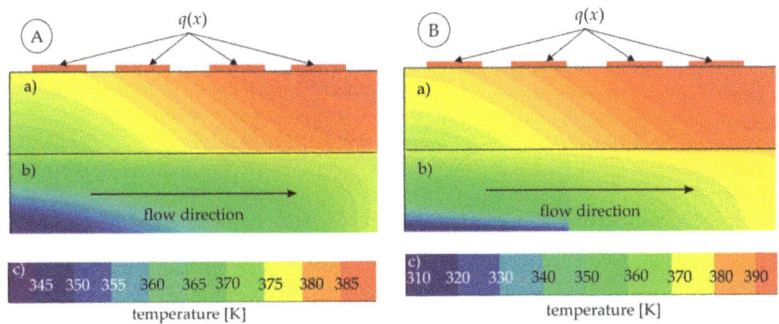

Figure 12. Two-dimensional temperature fields of (**a**) the copper block and (**b**) flowing fluid obtained by the Trefftz method for (**A**) heat flux $q = 223.2$ kW/m^2 and mass flux $G = 5.9$ kg/(m^2 s) and (**B**) heat flux $q = 340$ kW/m^2, mass flux $G = 5.9$ kg/(m^2 s), (pictorial view, not to scale).

Fluid flowing into the minichannel substantially reduces the temperature of the copper block across the contact area, Figure 12. The fluid motion in the minichannel causes heat transfer in the direction of its axis. As a result, the temperature in the entire channel increases as does the void fraction, which impairs heat transfer between the fluid and the heater. For this reason, in the final section of the module, both the liquid temperature and the heater temperature increase significantly.

The temperature distribution in both the fluid and the heating block, Figure 12, is as expected, the isotherms run in a manner likely for the direction and intensity of flow and location of the heaters. The charts are not fully comparable, because both parameters, G and q, are different for each case, but the impact of the parameter change on the temperature field is clear.

The results of numeric calculations presented in Figure 12 were obtained for experimental parameters given in Section 4. In the first step, nine T-complete functions were used to calculate the approximate two-dimensional temperature of the copper block T_C for Equation (5). Next, by combining four T-complete functions for Equation (20) and particular solution of Equation (11), the field of the fluid temperature T_f was found.

The boundary conditions adopted for the copper block and liquid temperatures were satisfied with high accuracy. The maximum differences between the calculated and measured temperatures of the copper block did not exceed 24 K (5.93%) for the first measuring point (from the inlet to the minichannel), 19 K (4.45%) for the second and 16 K (3.72%) for the third measurement point. Throughout the study, the largest errors were always at the first measuring point and the smallest at the last. When identifying the liquid temperature, the maximum differences between the calculated and measured temperature did not exceed 13 K (3.71%) at the inlet to the minichannel and 15 K (4.65%) at the outlet from the minichannel. As in the case of thermocouples placed in the copper block, the differences between the measured and calculated temperatures for liquids at the inlet to the minichannel were in most cases higher than the differences at the outlet from the minichannel.

Variation of the heat transfer coefficient for the flowing two-phase mixture is shown in Figure 13 as the function of the distance from the minichannel inlet. It is presented for three selected mass fluxes and varying heat flux. The heat transfer coefficient decreases as the distance from the inlet grows, which is related to increasing void fraction. Some variation in this trend can be observed both at the channel inlet and outlet. The growing heat flux also increases the heat transfer coefficient. Analysis of graphs in Figure 13 shows a known pattern—an increase in flow intensity causes an increase in heat transfer coefficient.

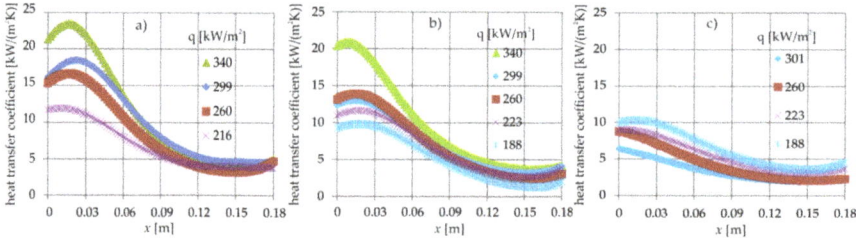

Figure 13. Heat transfer coefficient as a function of minichannel length (**a**) for G = 8.6 kg/(m^{2}s), (**b**) G = 5.9 kg/(m^2 s) and (**c**) G = 2.2 kg/(m^{2}s).

In the initial segment of the minichannel, changes in the calculated values of heat transfer coefficients as a function of distance from the inlet, Figure 13, are non-physical in nature. This is the result of boundary condition (6) adopted to stabilize the numeric procedure.

The shape of boiling curves presented in Figure 14 for various distances from the minichannel inlet is similar to the initial section of standard boiling curve. The curves originate with segments of steep dependence of the heat flux q versus the temperature difference $T_{C,s} - T_{sat}$ and flatten with further increase of that difference. This results from the growing vapor phase (void fraction) and approaching first boiling crisis. In the discussed experiment, the vapor was not superheated to prevent possible thermal damage of the research facility. Therefore, the boiling curves are limited to their initial segments.

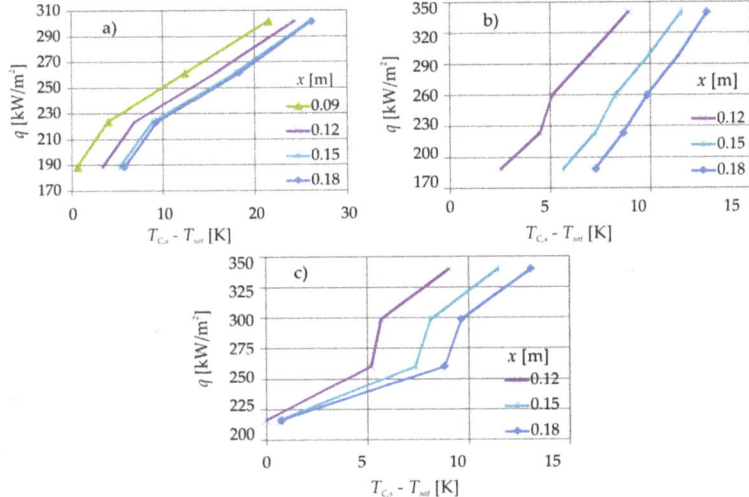

Figure 14. Boiling curves for selected points x along the minichannel length for (**a**) $G = 2.2$ kg/(m² s), (**b**) $G = 5.9$ kg/(m² s) and (**c**) $G = 8.6$ kg/(m² s).

The smaller the distance from the inlet, the greater is the heat flux q for the same temperature difference between the heating surface and the boiling liquid. This result agrees qualitatively with the results shown in Figure 13, because the heat transfer coefficient increases with the decreasing distance to the inlet.

With increasing distance from the channel inlet, the temperature difference $T_{C,s} - T_{sat}$ must increase to induce heat transfer of expected value. This is due to the increase in the amount of gaseous phase in the two-phase mixture and thus in its insulating properties.

The uncertainties of the experimental parameters determined in study [2] were applied to calculate the mean relative error (*MRE*) of the heat transfer coefficient $\alpha(x)$. Figure 15 shows the *MRE* as a function of both heat flux and mass flux. When the heat flux supplied to the copper block increases, the *MRE* decreases from the value of 4.2% for heat flux $q = 188$ kW/m² down to 1.0% for $q = 340$ kW/m². For smaller mass fluxes, the *MRE* values are lower.

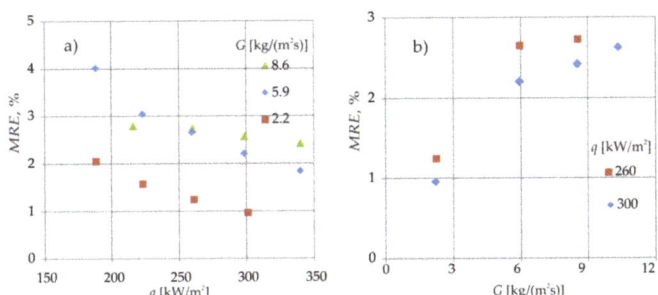

Figure 15. Mean relative error vs. (**a**) heat flux q, and (**b**) mass flux.

5. Conclusions

The experimental data gained at the Warsaw University of Technology, Plock Campus created a ground for the completed numeric analysis. The design of the experimental setup allowed setting and automatically acquiring temperature of the copper block and flowing fluid, fluid pressure, volumetric flow rate, heat flux and video recording for various positions of the video camera along the minichannel length. During the experimental run the research schedule of the setup was under control of the program created in the LabView environment. Data acquisition was controlled by the same script as well. The configuration of the setup is flexible enough to provide the potential for further expansion of the research program for new settings of thermal and flow parameters, channel space orientations and channel geometry.

The unique feature of the photographic method developed in this research is the video registration of the observed two-phase flow structure coupled with the measurement of the corresponding local value of the void fraction. This idea is based on the conversion of the flat image of the vapor bubble into the expected spatial shape. The obtained values of the void fraction were successfully applied in boiling two-phase flow modeling and numeric calculations. Disadvantages of the photographic method include 1) the requirement of quick collection of a large amount of data and 2) laborious conversion and refinement of the data into the required void fraction values with dedicated Matlab scripts.

The mathematical approach to solving two sequential IHCPs in flow boiling in the asymmetrically heated rectangular and horizontal minichannel was validated against experimental data. The employed meshless Trefftz method, making use of two types of T-complete functions, allowed obtaining approximate solutions of two coupled energy equations with their boundary conditions. The method provided stable solution of IHCP and consequently granted determination of: 1) the two-dimensional temperature distributions in two contacting domains, having different shapes and thermal features (copper block and flowing fluid), 2) the heat flux transferred from the block to the flowing two-phase mixture, and 3) the heat transfer coefficient on the contact surface between the copper block and flowing fluid. The Trefftz method is not limited by the number or character of boundary conditions. They can be established directly from temperatures or temperature gradients and they may have continuous or discrete form as well. It is worth noting that the approximate solutions satisfy the governing partial differential equations exactly and numeric calculations are not very much complicated because T-complete functions are polynomials in the investigated case.

In the near future the research program will be extended to include water, ethanol and FC72 and various minichannel space orientations [21].

Author Contributions: Conceptualization, S.H., M.E.P.; methodology, M.G, K.P., S.H. and B.M.; validation, S.H., B.M.; formal analysis, S.H.; investigation, M.G. and K.P.; writing—Original draft preparation, M.G., S.H. and M.E.P.; writing—Review & editing, S.H., B.M. and M.E.P.; funding acquisition, M.G. All authors have read and agreed to the published version of the manuscript.

Funding: The research was funded by the Warsaw University of Technology, Plock Campus, grant no. 504/04480/7193/44.000000 and the Poland National Science Center, grant no. UMO-2018/31/B/ST8/01199.

Conflicts of Interest: The authors declare no conflicts of interest.

Nomenclature

Nomenclature

A	area, cross-section, m^2
a	channel width, m
b	channel depth, m
c_p	specific heat, J/(kgK)
D	domain, mm
f	Fanning friction factor
G	mass flux, kg/(m^2 s)
H	height, m
L	length, m
MRE	mean relative error
p	pressure, Pa
P	ellipse semi-axis, m
Pr	Prandtl number

q	heat flux, kW/m²
R	radius, m
Re	Reynolds number
T	temperature, K, °C
w	velocity, m/s
x	coordinate in the flow direction, m
y	coordinate perpendicular to the flow direction, m
V	volume, m³
$\nabla^2 = \frac{\partial^2}{\partial x^2} + \frac{\partial^2}{\partial y^2}$	Laplacian in Cartesian coordinates

Greek symbol

α	heat transfer coefficient, W/(m² K)
Δ	difference
Φ	Rayleigh dissipation function, s⁻²
ϕ	void fraction
λ	thermal conductivity, W/(m K)
μ	dynamic viscosity, Pa s
ρ	density, kg/m³
Ω	negative heat source, W/m³

Subscripts

$approx$	approximation
ave	average
b	bubble
C	copper block
cam	observed part of the minichannel
con	convection
f	fluid
h	hydraulic
i	i–th bubble,
in	inlet
lb	large bubble
out	outlet
s	surface
sat	saturation
sb	small bubble
T	thermal
X	in the flow direction

References

1. Kim, S.-M.; Mudawar, I. Review of databases and predictive methods for heat transfer in condensing and boiling mini/micro-channel flows. *Int. J. Heat Mass Transf.* **2014**, *77*, 627–652. [CrossRef]
2. Grabowski, M.; Hożejowska, S.; Poniewski, M.E. Trefftz method-based identification of heat transfer coefficient and temperature fields in flow boiling in an asymmetrically heated rectangular mini-channel. In Proceedings of the XV Symposium on Heat and Mass Transfer 2019 (SOHAMT 2019), Kołobrzeg, Poland, 16–19 September 2019; Volume 1, pp. 179–192.
3. Karayiannis, T.G.; Mahmoud, M.M. Flow boiling in microchannels: Fundamentals and applications. *Appl. Therm. Eng.* **2017**, *115*, 1372–1397. [CrossRef]
4. Płaczkowski, K.; Poniewski, M.E.; Grabowski, M.; Alabrudziinski, S. Photographic technique application to the determination of void fraction in two-phase flow boiling in minichannels. *Appl. Mech. Mater.* **2015**, *797*, 299–306. [CrossRef]
5. Saisorn, S.; Kaew-On, J.; Wongwises, S. Flow pattern and heat transfer characteristics of R-134a refrigerant during flow boiling in a horizontal circular mini-channel. *Int. J. Heat Mass Transf.* **2010**, *53*, 4023–4038. [CrossRef]

6. Trefftz, E. Ein Gegenstück zum Ritzschen Verfahren. In Proceedings of the International Kongress für Technische Mechanik, Zürich, Switzerland, 12–17 September 1926; pp. 131–137.
7. Piasecka, M.; Hożejowska, S.; Poniewski, M.E. Experimental evaluation of flow boiling incipience of subcooled fluid in a narrow channel. *Int. J. Heat Fluid Flow* **2004**, *25*, 159–172. [CrossRef]
8. Hożejowska, S.; Piasecka, M.; Poniewski, M.E. Boiling heat transfer in vertical minichannels. Liquid crystal experiments and numerical investigations. *Int. J. Therm. Sci.* **2009**, *48*, 1049–1059. [CrossRef]
9. Hożejowska, S.; Kaniowski, R.; Poniewski, M.E. Trefftz method for calculating temperature field of the boiling liquid flowing in a minichannel. *Int. J. Numer. Methods Heat Fluid Flow* **2014**, *24*, 811–824. [CrossRef]
10. Grabowski, M.; Hożejowska, S.; Pawinska, A.; Poniewski, M.E.; Wernik, J. Heat Transfer Coefficient Identification in Mini-Channel Flow Boiling with the Hybrid Picard-Trefftz Method. *Energies* **2018**, *11*, 2057. [CrossRef]
11. Fan, C.-M.; Chan, H.-F.; Kuo, C.-L.; Yeih, W. Numerical solutions of boundary detection problems using modified collocation Trefftz method and exponentially convergent scalar homotopy algorithm. *Eng. Anal. Bound. Elem.* **2012**, *36*, 2–8. [CrossRef]
12. Płaczkowski, K. Application of Photographic Technology to the Two-Phase Flow Regime Recording and Concurrent Void Fraction Quantification (In Polish). Ph.D. Thesis, Warsaw University of Technology, Płock, Poland, 2020.
13. Hożejowska, S.; Grabowski, M.; Poniewski, M.E. Implementation of a three-dimensional model for the identification of flow boiling heat transfer coefficient in rectangular mini-channel. In Proceedings of the International Conference on Experimental Fluid Mechanics, Franzensbad, Czech Republic, 19–22 November 2019; Dančová, P., Ed.; Technical University of Liberec and MiTpp: Liberec, Czech Republic, 2001; pp. 178–182.
14. Bilicki, Z. Latent heat transport in forced boiling flow. *Int. J. Heat Mass Transf.* **1983**, *26*, 559–565. [CrossRef]
15. Bilicki, Z. The relation between the experiment and theory for nucleate forced boiling, Invited lecture. In Proceedings of the 4th ExHFT, Brussels, Belgium, 2–6 June 1997; pp. 571–578.
16. Bohdal, T. Modeling the process of bubble boiling on flows. *Arch. Thermodyn.* **2000**, *21*, 34–75.
17. Tolubinski, V.I.; Kostanchuk, D.M. Vapour bubbles growth rate and heat transfer intensity at subcooled water boiling. In Proceedings of the 4th International Heat Transfer Conference, Paris-Versailles, France, 31 August–5 September 1970; pp. 1–11.
18. Labuncov, D.A. Obobščennyie zavisimosti dla teplootdači pri puzirkovom kipenii židkosti. *Teploenergetika* **1960**, *5*, 76–81.
19. Shah, R.K.; London, A.L. *Laminar Flow Forced Convection in Ducts*; Academic Press: New York, NY, USA, 1978.
20. Hożejowski, L.; Hożejowska, S. Trefftz method in an inverse problem of two-phase flow boiling in a minichannel. *Eng. Anal. Bound. Elem.* **2019**, *98*, 27–34. [CrossRef]
21. Layssac, T.; Lips, S.; Revellin, R. Effect of inclination on heat transfer coefficient during flow boiling in a mini-channel. *Int. J. Heat Mass Transf.* **2019**, *132*, 508–518. [CrossRef]

© 2020 by the authors. Licensee MDPI, Basel, Switzerland. This article is an open access article distributed under the terms and conditions of the Creative Commons Attribution (CC BY) license (http://creativecommons.org/licenses/by/4.0/).

Article

A Study on the Application Possibility of the Vehicle Air Conditioning System Using Vortex Tube

Younghyeon Kim [1], Seokyeon Im [2,*] and Jaeyoung Han [3,*]

[1] Department of Mechanical Engineering, Chungnam National University, 99 Daehak-ro, Yuseong-gu, Daejeon 34134, Korea; viny9198@naver.com
[2] Department of Mechanical and Materials Engineering Education, Chungnam National University, 99 Daehak-ro, Daejeon 34134, Korea
[3] School of Mechanical & Automotive Engineering, Youngsan University, 288 Junam-ro, Yangsan-si 50510, Korea
* Correspondence: imsy@cnu.ac.kr (S.I.); hjyt11@ysu.ac.kr (J.H.); Tel.: +82-42-821-7992 (S.I.); +82-055-380-9473 (J.H.); Fax: +82-42-821-8732 (S.I.); +82-55-380-9249 (J.H.)

Received: 9 August 2020; Accepted: 7 October 2020; Published: 8 October 2020

Abstract: Since refrigerants applied to vehicle air conditioning systems exacerbate global warming, many studies have been conducted to supplement them. However, most studies have attempted to maximize the efficiency and minimize the environmental impact of the refrigerant, and thus, an air conditioning system without refrigerant is required. The vortex tube is a temperature separation system capable of separating air at low and high temperatures using compressed air. When applied to an air conditioning system, it is possible to construct an eco-friendly system that does not use a refrigerant. In this paper, various temperature changes and characteristics of a vortex tube were identified and applied to an air conditioning system simulation device. Additionally, an air conditioning system simulation device using indirect heat exchange and direct heat exchange methods was constructed to test the low-temperature air flow rate (y_c), according to the temperature and pressure. As a result of the experiment, the temperature of the indirect heat exchange method was found to be higher than the direct heat exchange method, but the direct heat exchange method had low flow resistance. As a result, the direct heat exchange method can easily control the temperature according to the pressure and the low-temperature air flow rate (y_c). Therefore, it was judged that the direct heat exchange method is more feasible for use in air conditioning systems than the indirect heat exchange method.

Keywords: vortex generator; vortex tube; temperature separation; the low-temperature air flow ratio (y_c), inlet pressure (P_i)

1. Introduction

With rapid industrialization, the share of fossil fuels has increased significantly from the beginning of the industrial revolution to the present, and as means of transportation and efficient heating and cooling systems have been developed, more environmental pollution has been produced. Furthermore, the excessive use of refrigerant has caused environmental problems such as ozone layer destruction and global warming, and the ripple effects are accelerating. Accordingly, restrictions on the use of CFC (Chloro Fluoro Carbon)-based refrigerants have been strengthened, and R134a has been developed as an alternative. This refrigerant has been applied to facilities such as automobile air conditioning systems and household refrigerators. However, while R134a has an ozone depletion index of 0, its global warming index of 1430 is insufficient to compensate for environmental problems [1,2]. As an alternative to solve the problems associated with freon (CFC) used as a refrigerant in automobiles, research has been actively conducted, where one possible measure is the vortex tube [3].

The vortex tube was discovered by Ranque in 1928 and has since become widely promoted to the academic community by Hilsch [4,5]. Thereafter, modeling studies were conducted by Fulton, Stephan, Deissler and Permuter, Kassner, Upendra, and Gao et al. for energy separation, and models proposed by Fulton and Kassner have been widely accepted [6–11]. Hilsch et al. experimented to find nozzle diameters and low-temperature exit conditions for three cases of tube radii of 4.6 mm, 9.6 mm, and 17.6 mm. Westley et al. conducted a study on vortex tube dimensional conditions to find the ratio of the tube area of the low-temperature air outlet area to the tube area of the nozzle's total area to spray compressed air into the tube [12]. Hartnett and Eckert et al. measured the tube velocity, pressure, and temperature distribution through experiments and found that the vortex tube had a high wall temperature and a low central temperature [13]. Takanama et al. designed the shape of a vortex tube with high energy efficiency by researching the relationship between the temperature and velocity distribution of air flowing in the vortex internal production chamber and the dimensions of the tube, nozzle, and low-temperature outlet orifice [14]. Comassar and Stepan et al. presented a single-flow vortex tube to evaluate the energy separation performance and confirmed that the counter-flow vortex tube performance was higher than the single-flow vortex tube [15,16]. Frohlingsdorf et al. analyzed energy separation and flow in a Ranque–Hilsch vortex tube using CFX, shear stress, and axial symmetry models [17]. Smith Eiamsa-ard et al. performed a thermal separation simulation of a Ranque–Hilsch vortex tube in a finite volume approach using a standard turbulence model and a logarithmic stress model (ASM) [18]. Skye et al. compared the performance predicted by the CFD model with commercial vortex tube experimental data and confirmed the industrial applicability of vortex tubes [19].

Until recently, studies on the industrial use of vortex tubes are ongoing, and experimental analysis of vortex tube energy separation phenomena and comparative analysis through models are continuously being studied [20–23]. B.E. Mtopi Fotso et al., and others, showed modeling and thermal analysis of solar generators equipped with vortex tubes for hybrid vehicles, showing the application of vortex tubes to vehicles [24]. Additionally, Hitesh R. Thakare et al. performed a technical and economic evaluation of the vortex tube's industrial applications to show that the industrial use of vortex tubes is advantageous for shortening production cycles [25].

In this study, the vortex tube was found to be eco-friendly because it does not require a refrigerant; and because it is sufficiently applicable to air conditioning systems in the future automobile field, a study was conducted to integrate a vortex tube into a vehicle air conditioning system. The experiment was performed by dividing it into two parts: an indirect heat exchange method and a direct heat exchange method. In particular, among the factors affecting the energy separation of the vortex tube, the temperature is controlled by controlling the supply pressure (P_i), the nozzle area ratio (S_n), and the number of nozzle holes (N_h) of the vortex tube generator, and the low-temperature air flow rate (y_c) is controlled using a throttle valve. We obtained the experimental data for the four basic and influential characteristics in the characteristic experiment, and based on this, analyzed the optimum temperature characteristics by installing them in heat exchangers of different shapes using the indirect and direct methods.

2. Experimental Approach

2.1. Principle of Vortex Tube

The vortex tube is an energy separation device that separates compressed air into hot and cold air samples without chemical reaction or combustion. The vortex tube has a simple structure, but the cause and phenomenon of energy separation are complicated. In general, Fulton's theory describing the domains of free vortex and forced vortex is accepted. According to the theory, compressed fluid forms a free vortex with high velocity at the tangential nozzle of the tube.

$$wr^2 = const. \tag{1}$$

At this time, as the flow proceeds toward the throttle valve, the free vortex is the result of friction between the fluid layers, and a forced vortex is formed in the center of the tube.

$$w = const. \tag{2}$$

Due to this change in flow type, momentum transfer occurs from the center of the tube to the wall, and the center of the vortex is cooled more than the outside. However, to achieve energy balance, heat is transferred to the inner layer; however, since the momentum transfer is greater than the transfer of heat, the outside air temperature rises as does the stagnant temperature, and air is discharged through the throttle valve as a high-temperature flow. At this time, the fluid inside drops to a lower temperature by losing momentum greater than the received thermal energy, and becomes a low-temperature flow with temperature lower than the inlet temperature; this air is discharged to the outside through the low-temperature outlet orifice. Figures 1 and 2 clearly show the phenomenon of this vortex tube [26]. What is shown in the vertical direction in Figure 1 is a visual representation of the force magnitude in the vortex tangent direction.

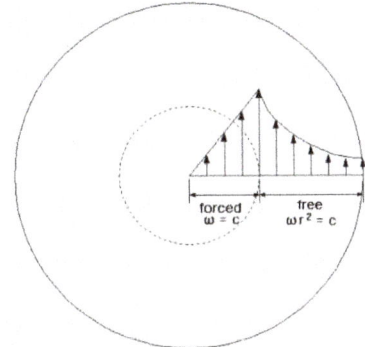

Figure 1. Cross-section of the vortex tube in the "free" and "forced" vortex flow.

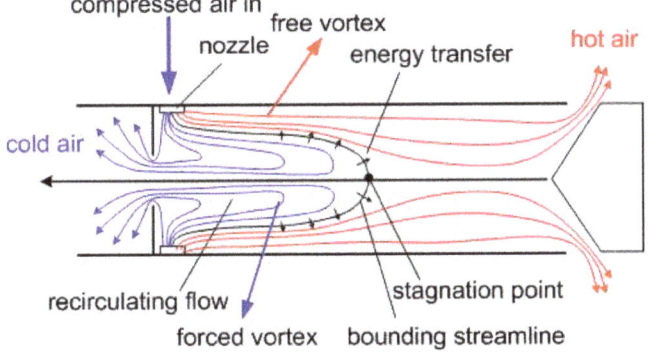

Figure 2. Supposed flow pattern and energy flow in the counter flow type.

2.2. Setup of Feasibility Study

In this study, the vortex tube device was applied to a vehicle air conditioning system. Therefore, through a basic experiment with an actual manufactured vortex tube, it was confirmed whether energy separation had been properly performed; also, the number of nozzle holes (N_h) of the vortex generator was changed according to the supply pressure (P_i) to change the temperature according to the low-temperature airflow ratio (y_c). Schematics of the basic experimental device and the actual experimental device are shown in Figures 3 and 4.

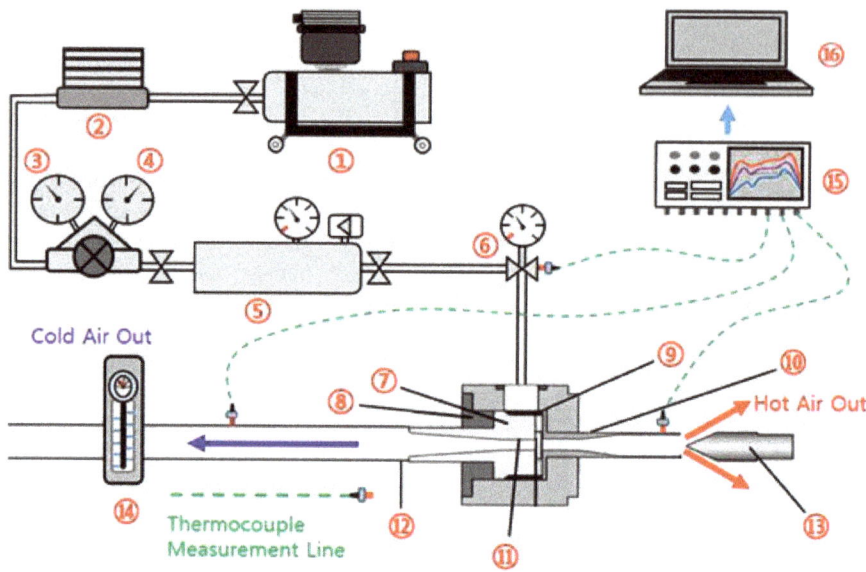

① Compressor	② Air Dryer	③ Pressure regulator	④ Air filter
⑤ Air surge	⑥ Check valves	⑦ Vortex generator	⑧ Holder
⑨ Nozzle	⑩ Sleeve	⑪ Orifice	⑫ Vortex tube
⑬ Throttle valve	⑭ Air flowmeter	⑮ Data logger	⑯ Lap top

Figure 3. Schematic diagram of the vortex tube system.

The specifications of the vortex tube produced in the experiment are shown in Table 1. The air conditioning system to be applied to the experiment was divided into two types: an indirect heat exchange method and a direct heat exchange method.

Table 1. Dimension of the vortex tube.

Parameter	Value	Unit
Tube length (L)	128	mm
Inlet diameter of vortex tube (D)	9.8	mm
Inner diameter of nozzles (D_n)	1, 1.2, 1.4, 1.7, 2.1	mm
Diameter of cold end orifice (d_c)	6	mm
Number of nozzle holes (N_h)	4, 5, 6, 7, 8	-
Nozzle area ratio (S_n)	0.142	-

Compressed air that has undergone energy separation through the vortex tube first controls the temperature through a low-temperature and high-temperature valve, and this air is moved into the chamber through an indirect heater core where heat exchange is performed using a direct air conditioning filter. The temperature is measured. The heater core of the indirect heat exchanger used in the experiment is shown in Figure 5; the air conditioning filter of the direct heat exchanger is shown in Figure 6.

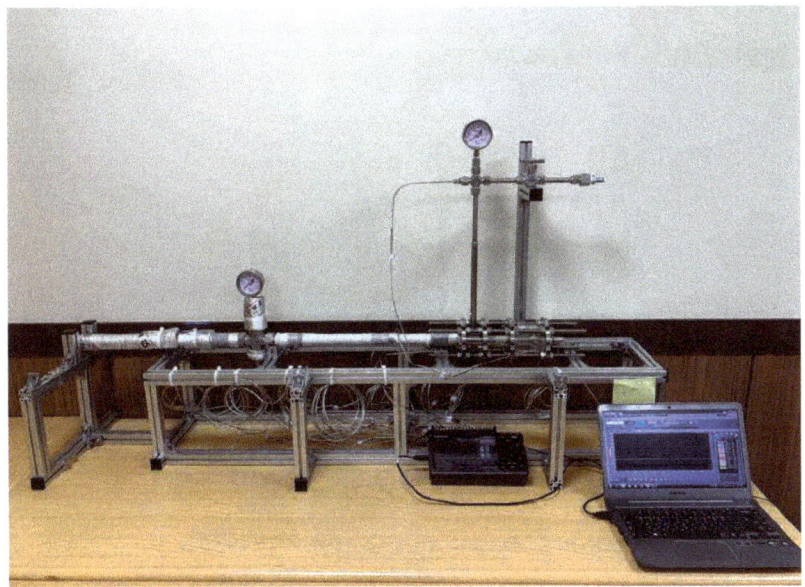

Figure 4. Photograph of the experimental apparatus for the vortex tube.

Figure 5. Indirect type heater core.

Figure 6. Direct type air conditioning filter.

2.3. Experimental Apparatus

Figures 7 and 8 show the schematic and actual experimental device for the indirect heat exchanger. Figures 9 and 10 show a schematic of the direct heat exchange method and an actual experimental device. Configurations of the experimental device for the vortex tube can be largely divided into three types: a supply unit that supplies compressed air, a test unit that performs temperature separation experiments, and a control unit that controls the system. The supply section consists of an air compressor, air dryer, air surge tank, and pressure regulator, which supplies compressed air. The test section consists of a vortex tube and a throttle valve, thermocouple, pressure gauge, flow meter, etc., that can control the low-temperature air flow rate (y_c). The control part is composed of a part that controls pressure and flow rate and a part that checks data from each temperature sensor. The air entering the inlets (shown in Figures 7 and 9) is hot and cold air separated through the vortex tube; this air is supplied to each inlet. Table 2 shows the specifications of the device used in the experiment.

The compressed air produced by the compressor enters the surge tank. After enough air is present in the surge tank, it passes through the filter and is supplied to the vortex tube inlet. This phenomenon is a process that is conducted to catch the pulsation of the air generated by the compressor, and the following method was used to reduce the experimental error. Then, the supplied air passes through the vortex generator, and the vortex is generated inside the experimental apparatus. The generated vortex induces an energy separation phenomenon in the tube structure, and air separates into hot air and cold air and is discharged. Each type of air is then heat exchanged through the heater core. The air that has undergone heat exchange and the outside air are mixed in the chamber; according to the flow of the air, the experiment was performed by dividing it into indirect and direct methods. If there was no difference of 0.1 °C for 100 s, the temperature was determined to be in a normal state, and the corresponding temperature was measured. In addition, the reliability of the experimental results was secured through repeated experiments.

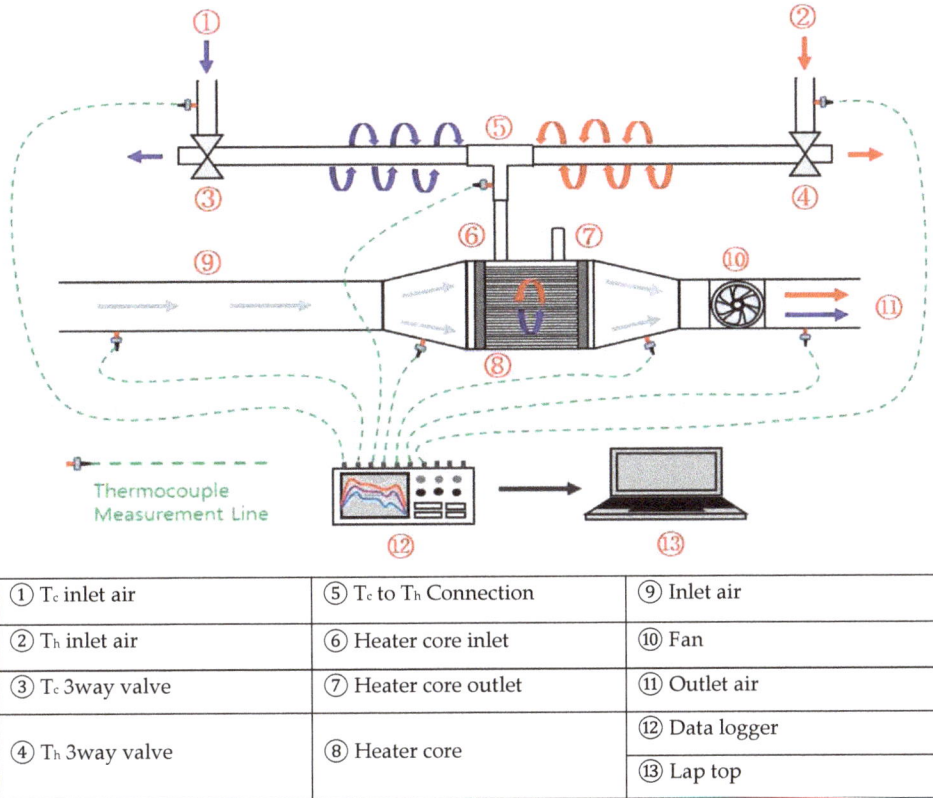

① T_c inlet air	⑤ T_c to T_h Connection	⑨ Inlet air
② T_h inlet air	⑥ Heater core inlet	⑩ Fan
③ T_c 3way valve	⑦ Heater core outlet	⑪ Outlet air
④ T_h 3way valve	⑧ Heater core	⑫ Data logger
		⑬ Lap top

Figure 7. Schematic diagram of the indirect heater core.

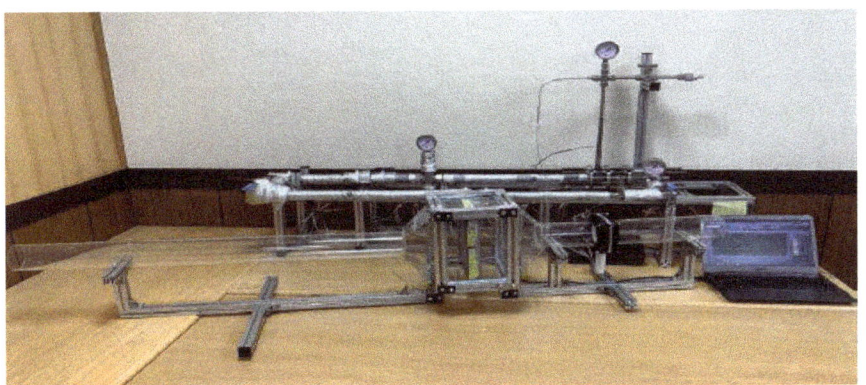

Figure 8. Photograph of an indirect experimental device with a heater core.

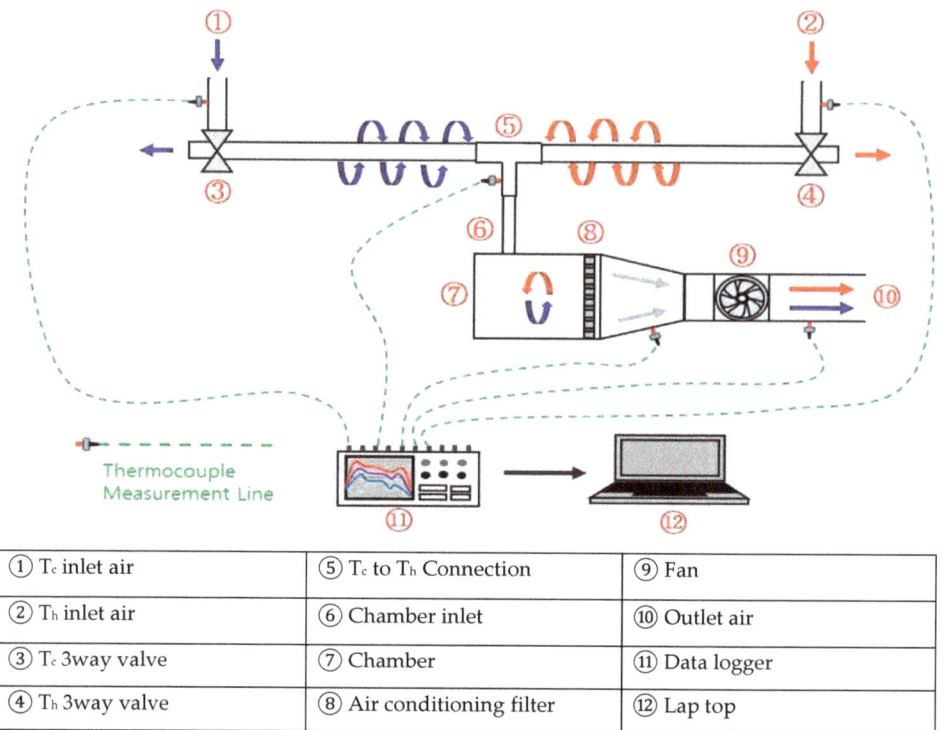

① T$_c$ inlet air	⑤ T$_c$ to T$_h$ Connection	⑨ Fan
② T$_h$ inlet air	⑥ Chamber inlet	⑩ Outlet air
③ T$_c$ 3way valve	⑦ Chamber	⑪ Data logger
④ T$_h$ 3way valve	⑧ Air conditioning filter	⑫ Lap top

Figure 9. Schematic diagram of the direct air conditioning filter.

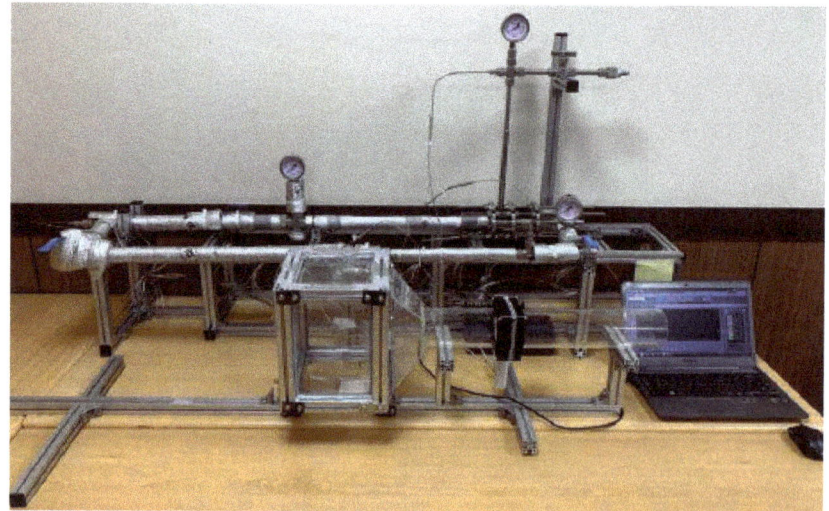

Figure 10. Photograph of the direct experimental device with air conditioning filter.

Table 2. Range and uncertainly of the equipment used in the experiment.

Instrument	Model	Range	Uncertainly
Graphtec	GL 840 Midi Logger	-	±0.5%
Thermo-couple	K-type	−100 to 650 °C	±2.5 °C
Pressure sensor	Dream31 (A-type) Dream	−1 to 6 bar	±1.5%
Gas regulator	Renown	0–6 bar	±2.0%
Air compressor	Renown	20 HP	-
Air surge tank	Renown	280 L	-

3. Results and Discussion

3.1. Temperature Change According to Pressure and Generator Nozzle

The temperature change according to the pressure in the vortex tube device and the number of nozzles of the generator was investigated. The nozzle area ratio of $S_n = 0.142$ was set using a commonly known standard. The nozzle area ratio is the standard of the vortex tube that is set to best separate high temperature air. In addition, since the vortex formed in the vortex tube varies depending on the inlet supply pressure, the experiment was performed between $P_i = 0.5$ kgf/cm^2 and $P_i = 5.0$ kgf/cm^2 at $P_i = 0.5$ kgf/cm^2 intervals. For each experiment, the high-temperature air temperature difference ΔT_h and the low-temperature air temperature difference ΔT_c at low temperature air flow rate (y_c) = 0.1~1.0 were measured. Results are shown in Figures 11–20.

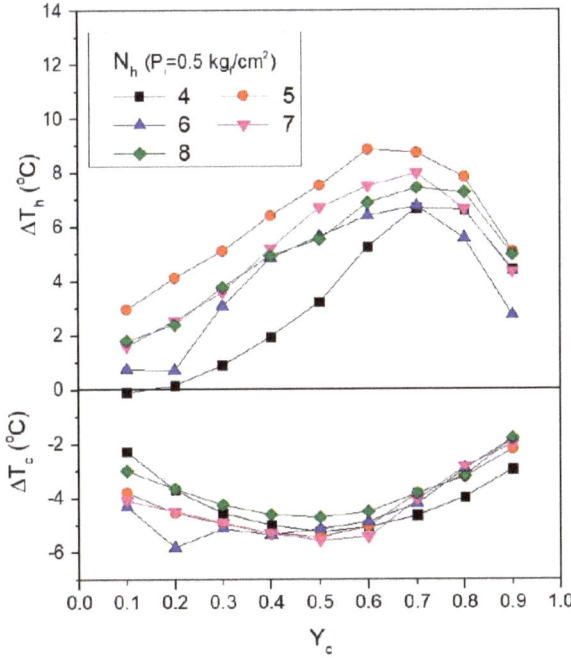

Figure 11. Temperature differences ΔT_h and ΔT_c according to inlet pressure (P_i) 0.5 kgf/cm^2.

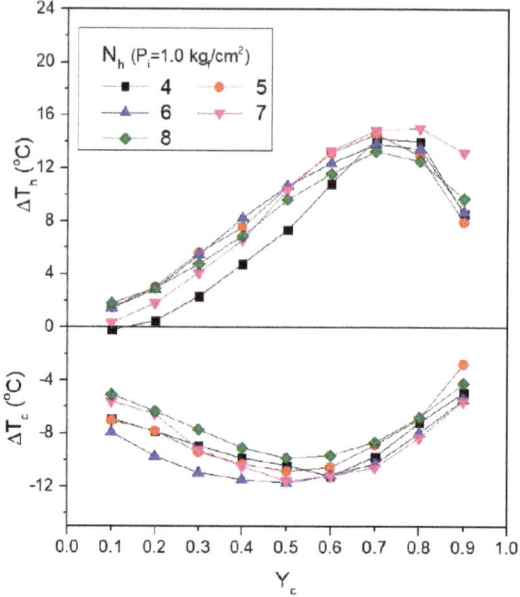

Figure 12. Temperature differences ΔT_h and ΔT_c according to inlet pressure (P_i) 1.0 kgf/cm^2.

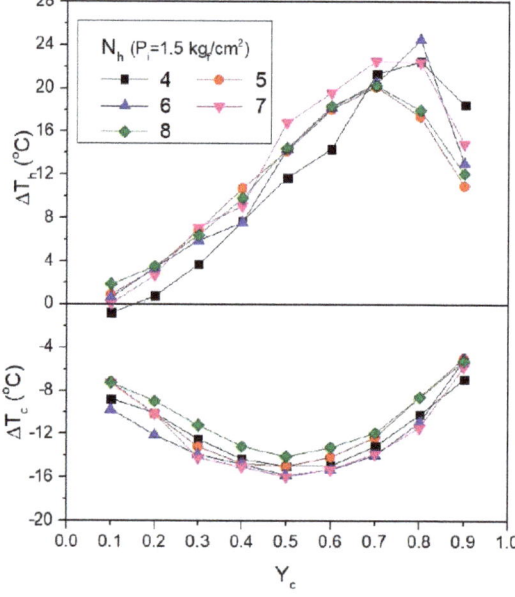

Figure 13. Temperature differences ΔT_h and ΔT_c according to inlet pressure (P_i) 1.5 kgf/cm^2.

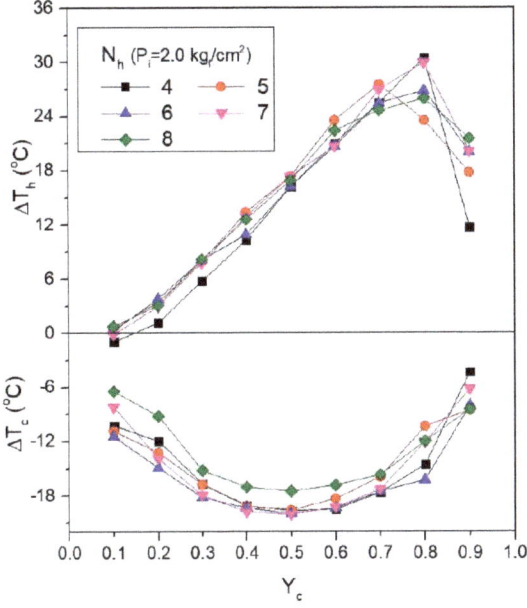

Figure 14. Temperature differences ΔT_h and ΔT_c according to inlet pressure (P_i) 2.0 kgf/cm^2.

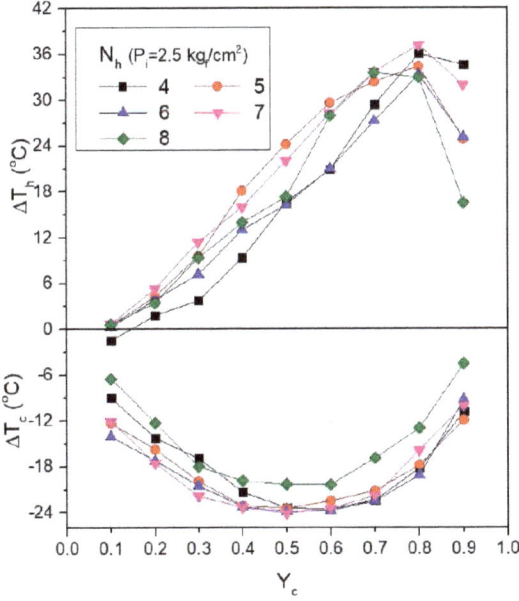

Figure 15. Temperature differences ΔT_h and ΔT_c according to inlet pressure (P_i) 2.5 kgf/cm^2.

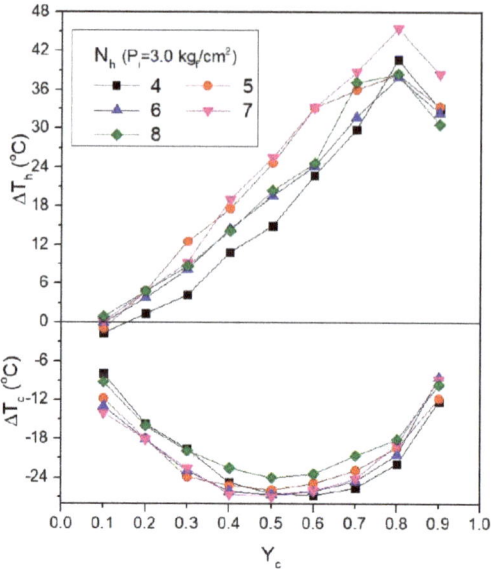

Figure 16. Temperature differences ΔT_h and ΔT_c according to inlet pressure (P_i) 3.0 kgf/cm^2.

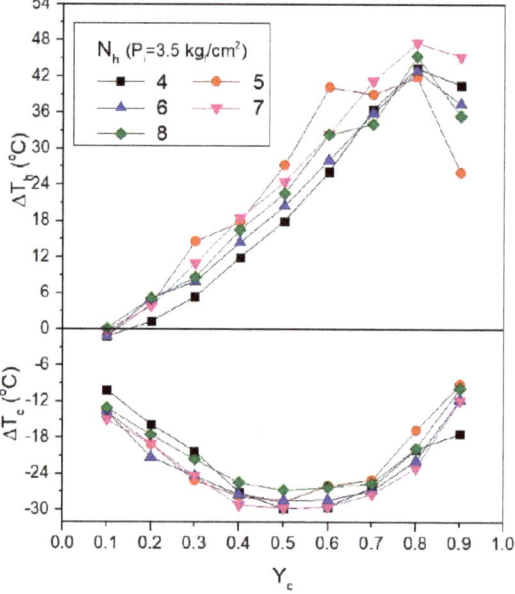

Figure 17. Temperature differences ΔT_h and ΔT_c according to inlet pressure (P_i) 3.5 kgf/cm^2.

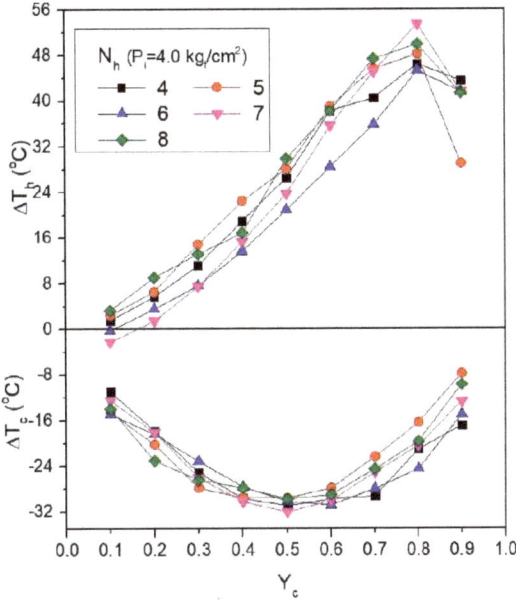

Figure 18. Temperature differences ΔT_h and ΔT_c according to inlet pressure (P_i) 4.0 kgf/cm^2.

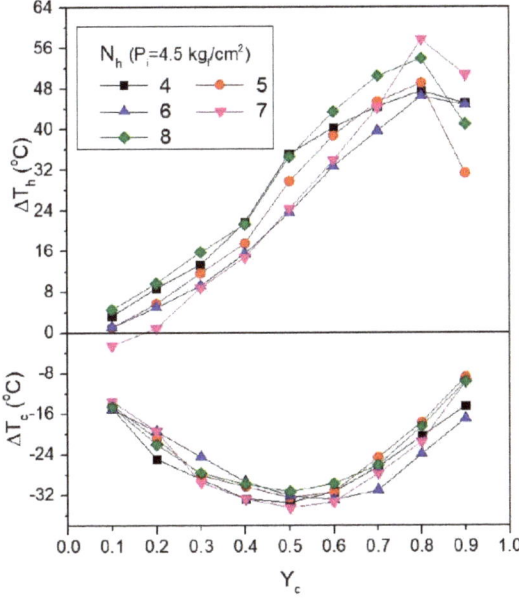

Figure 19. Temperature differences ΔT_h and ΔT_c according to inlet pressure (P_i) 4.5 kgf/cm^2.

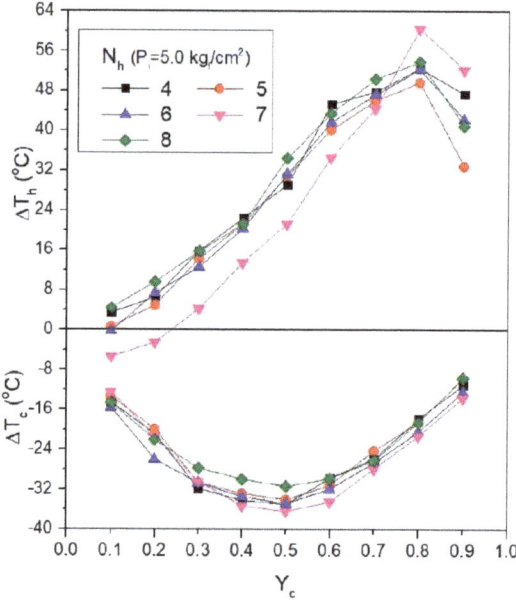

Figure 20. Temperature differences ΔT_h and ΔT_c according to inlet pressure (P_i) 5.0 kgf/cm².

According to the experimental results, it was confirmed that ΔT_h showed the highest high-temperature air from y_c = 0.6 to 0.8 at high temperatures, and y_c = 0.8 in the case of $\Delta T_{h,\,max}$ as the pressure increased. In the case of low temperature, ΔT_c showed the maximum low-temperature air from y_c = 0.4 to 0.6, and in the case of $\Delta T_{c,\,max}$ in the same case as the high temperature, y_c = 0.5. As a result of these experiments, when the throttle valve on the high-temperature outlet is closed, y_c increases and the flow cross-sectional area of the high-temperature outlet decreases. Therefore, the air at the high-temperature outlet flows back to the relatively low-temperature outlet, and the central flow becomes smooth. Additionally, when y_c relatively decreases, the flow cross-sectional area of the high-temperature outlet becomes wider, and ΔT_c tends to increase because the stagnation point in the vortex tube moves more toward the high-temperature outlet than it does when y_c increases. It was judged that ΔT_h and ΔT_c increased because of active energy separation due to increases in momentum transfer between the central flow and the flow near the wall due to the change in y_c. Through the experiments, it was found that ΔT_h and ΔT_c increased in a similar manner as the inlet pressure (P_i) flowing into the vortex tube increased. As the inlet pressure increased, the flow rate increased, and so the energy loss of the flow decreased, regardless of the number of nozzles, and the energy separation phenomenon became relatively large. This experiment trend showed similar tendencies of low-temperature outlet flow rate and entire pressure section, just as in Stephan [7]. Table 3 shows the $\Delta T_{h,\,max}$ and $\Delta T_{c,\,max}$ values of y_c, and the number of nozzles depending on the pressure; in the case of T_h, it can be seen that the number of nozzles varied from inlet pressure P_i = 0.5 to 2.0 kgf/cm². However, after P_i = 2.0 kgf/cm², it can be seen that the number of nozzles appears constant; in the case of T_c, the number of nozzles seems to change at P_i = 1.0 kgf/cm². Overall, in the case of T_h, the temperature was highest at y_c = 0.8, and the nozzles were found to be efficient when they had seven holes. T_c was found to have its lowest temperature at y_c = 0.5, and the number of nozzles was found to be efficient at seven.

Table 3. $T_{h,\,max}$ and $T_{c,\,max}$ of y_c and nozzle holes depending on the pressure.

P_i	$T_{h,\,max}$	y_c	N_h	$T_{c,\,max}$	y_c	N_h
0.5	8.86 °C	0.6	5 hole	−5.57 °C	0.5	7 hole
1.0	15.06 °C	0.8	7 hole	−14.17 °C	0.5	6 hole
1.5	24.51 °C	0.8	6 hole	−15.90 °C	0.5	7 hole
2.0	30.42 °C	0.8	4 hole	−20.08 °C	0.5	7 hole
2.5	37.17 °C	0.8	7 hole	−24.06 °C	0.5	7 hole
3.0	45.55 °C	0.8	7 hole	−26.90 °C	0.5	7 hole
3.5	47.61 °C	0.8	7 hole	−29.66 °C	0.5	7 hole
4.0	53.38 °C	0.8	7 hole	−32.05 °C	0.5	7 hole
4.5	57.70 °C	0.8	7 hole	−34.46 °C	0.5	7 hole
5.0	60.51 °C	0.8	7 hole	−36.33 °C	0.5	7 hole

3.2. Discharge Flow Rate and Temperature Difference between $\Delta T_{h,\,max}$ and $\Delta T_{c,\,max}$ According to Pressure

As an initial experiment, we tried to determine the total flow rate according to the supply pressure. The table that summarizes the data according to the number of nozzles is shown in Figure 21.

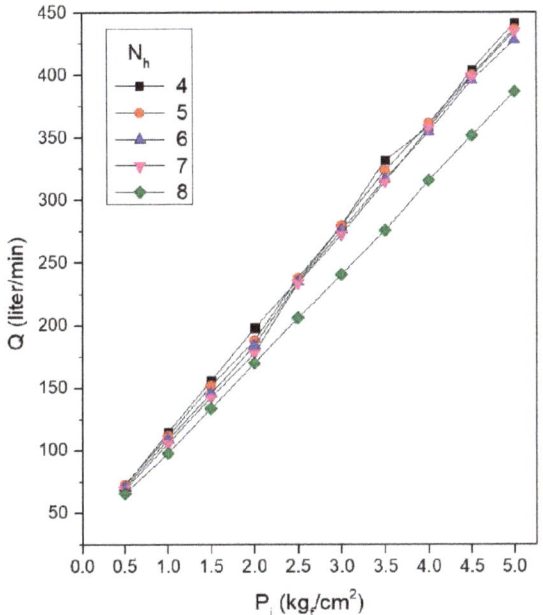

Figure 21. Total discharge flow rate of intake air pressure by the number of nozzles.

The nozzle cross-section (S_n) of the generator was selected as a high-temperature type. As a result of experiments on low-temperature and high-temperature type nozzles, the temperature difference was found not to be large in the low-temperature region, but showed a difference of about 8 °C in the high-temperature region. This phenomenon is considered to be due to an increase in flow resistance because the cross-sectional area of the high-temperature nozzle has a smaller area ratio than that of the low-temperature nozzle. This can be seen in the nozzle cross-section experiment [27]. The purpose of this experiment was to determine the conditions for effective temperature separation characteristics with minimal flow. Like in the basic experiment, the supply pressure between $P_i = 0.5$ kgf/cm² and $P_i = 5.0$ kgf/cm² to the vortex generator was set at the interval of $P_i = 0.5$ kgf/cm². In the experiment, it can be seen that the discharge flow rate increased proportionally as the inlet supply pressure increased, regardless of the number of generator nozzles. Under the same pressure condition, the number of nozzles tends to slightly increase as the size of the nozzle decreases, which means that when the generator's nozzle cross-section (S_n) is the same, as the number of generator nozzles decreases, the nozzle inner diameter (D_n) of the nozzle cross-section (S_n) increases. Therefore, it is considered that the total discharge flow rate increases because the flow resistance of compressed air, through which the supply pressure flows through the generator nozzle, decreases. From the experimental results, according to the number of nozzles and the supply pressure of the vortex generator, the maximum high-temperature air temperature difference ($\Delta T_{h,\,max}$) and the maximum low-temperature air temperature difference ($\Delta T_{c,\,max}$) under each condition are determined and shown in Figure 22.

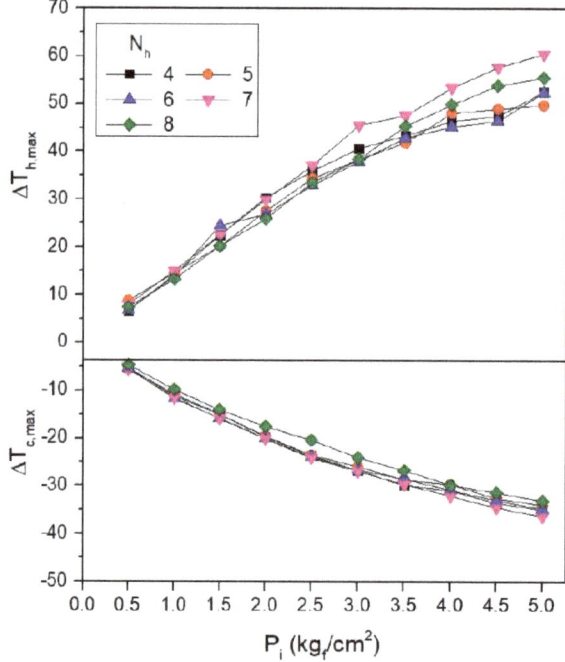

Figure 22. Temperature difference $\Delta T_{h,\,max}$ and $\Delta T_{c,\,max}$ according to the number of nozzle holes.

As a result of the experiment, most high-temperature and low-temperature air temperature separation characteristics showed good efficiency under certain conditions in the case of the seven nozzles, but there was not a large difference in temperature. Furthermore, it can be seen that for the temperature separation characteristic, the maximum high-temperature air temperature difference ($\Delta T_{h,\,max}$) and the maximum low-temperature air temperature difference ($\Delta T_{c,\,max}$) changed significantly as the supply pressure increased. The high-temperature air of the seven nozzles with high efficiency changed from 7.99 °C when the maximum hot air temperature difference ($\Delta T_{h,\,max}$) $P_i = 0.5$ kgf/cm^2 to 60.51 °C when $P_i = 5.0$ kgf/cm^2, and showed a temperature difference of 52.52 °C. The low temperature air varied from −5.57 °C when the maximum low temperature air temperature difference ($\Delta T_{c,max}$) $P_i = 0.5$ kgf/cm^2 to −36.33 °C when $P_i = 5.0$ kgf/cm^2, and showed a temperature difference of 30.76 °C. The result of this experiment provides information for the application of a vehicle air conditioning system. Current air conditioning systems for vehicles have values of $P_i = 2.5$ kgf/cm^2 to $P_i = 5.0$ kgf/cm^2, where the temperature change is constant in a vehicle with an air conditioning system using a refrigerant. Based on the experimental results, it is considered that controlling the flow rate and temperature values required for heating and cooling by checking the discharge flow rate and the temperature difference between the hot and cold parts is a key point.

3.3. Temperature Control of Indirect Heat Exchange Method According to Pressure

To measure the temperature of the indirect heat exchanger, between $P_i = 0.5$ kgf/cm^2 and $P_i = 5.0$ kgf/cm^2 and $P_i =$ kgf/cm^2 were set based on the number of nozzles of the vortex generator, which was the most efficient in the basic experiment. In the high-temperature type device, the experiment was conducted at a section $y_c = 0.6$ to 0.8 at a point where the efficiency of the low-temperature air flow rate (y_c) was good; the experiment was conducted at $y_c = 0.4$ to 0.6 for a section of low-temperature type. Figure 23 shows these results graphically. As in the basic experiment, it can be seen that the vortex tube increases in efficiency with increasing pressure. In the experiment, the value of T_{aoc} (°C) is the point at which the temperature of both the vortex tube cold and hot outlets passes through the heater core. Additionally, the value of ΔT_{ac} (°C) was expressed by measuring the temperature of the flow flowing into the front and rear ends of the chamber surrounding the heater core. In the case of T_{aoc} (°C), the temperature was measured by comparing the low-temperature air flow rate (y_c) of the flow. In the low-temperature region, the temperature difference of T_{aoc} (°C) was not large, but it was 21.84 °C when $P_i = 0.5$ kgf/cm^2 at $y_c = 0.6$, where the efficiency was best in proportion to the discharge flow rate, and when $P_i = 5.0$ kgf/cm^2. The temperature difference changed to 9.87 °C and showed a maximum temperature difference of 11.97 °C. The value of ΔT_{ac} (°C) was also measured using the low-temperature airflow ratio (y_c), and the temperature difference was similar to T_{aoc} (°C). The value of ΔT_{ac} (°C) showed values of −0.03 °C when $P_i = 0.5$ kgf/cm^2 at $y_c = 0.6$, and −11.92 °C when $P_i = 5.0$ kgf/cm^2, resulting in a maximum temperature difference of 11.95 °C. Through this, the low-temperature measurement of the indirect heat exchange method according to the pressure, $y_c = 0.5$ was excellent in the basic experiment, but when heat exchange was performed, the resulting value of $y_c = 0.6$, which indicates a relatively large discharge flow rate due to flow rate resistance, was the best.

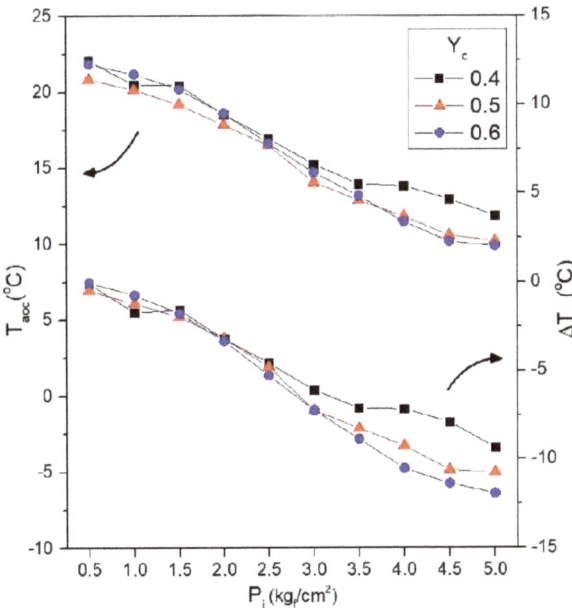

Figure 23. Indirect heat exchange method of ΔT_{aoc} and ΔT_{ac} according to pressure.

As can be seen in Figure 24, the temperature difference was higher in the high-temperature region than in the low-temperature region. At y_c = 0.8, the highest temperature of T_{aoc} (°C), when P_i = 0.5 kgf/cm², was 23.00 °C; when P_i = 5.0 kgf/cm², this value changed to 38.79 °C, showing a maximum temperature difference of 15.79 °C. The value of ΔT_{ac} (°C) was measured through the low-temperature airflow ratio (y_c), and the temperature difference had a result somewhat similar to that for T_{aoc} (°C). At y_c = 0.8, P_i = 0.5 kgf/cm² showed a difference of 0.02 °C, and P_i = 5.0 kgf/cm² showed a difference of 15.95 °C; the pressure difference resulted in a maximum temperature difference of 15.93 °C. However, in the high-temperature region of the indirect heat exchange method, it was confirmed that, unlike in the case at low temperature, the influence of pressure was large. In the case of y_c = 0.8, it can be seen that the discharge flow rate is low and the temperature change is small until the pressure becomes P_i = 3.0 kgf/cm²; the pressure changes significantly after P_i = 3.0 kgf/cm². Through these experiments, it is thought that the reaction rate of the heat exchanger will be higher than between y_c = 0.7 and 0.8 in the indirect heat exchange method when y_c = 0.6, which has a constant temperature change and a large discharge flow rate. Therefore, as a result of confirming the temperature difference according to the low-temperature air flow rate (y_c) through an indirect heat exchange method that changes according to the pressure, it was determined that a value of y_c = 0.6 is suitable for regions of both low temperature and high temperature.

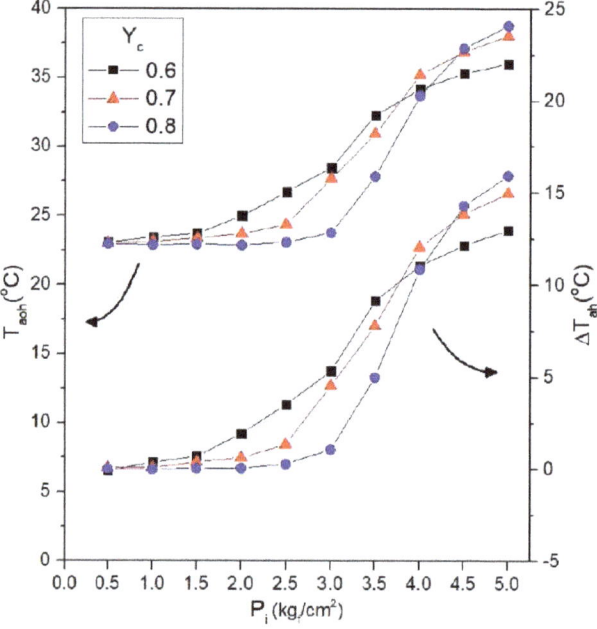

Figure 24. Indirect heat exchange method of ΔT_h according to pressure.

3.4. Temperature Control of Direct Heat Exchange Method According to Pressure

To evaluate the temperature difference with the indirect heat exchange method using the vortex tube, the supply pressure between P_i = 0.5 kgf/cm² and P_i = 5.0 kgf/cm² was the same as that in the basic experiment using seven nozzles for the vortex generator under the same test conditions. The inlet pressure was set at P_i = 0.5 kgf/cm² intervals. Furthermore, as the direct heat exchange method has a larger discharge flow rate than that of the indirect heat exchange method, experiments were conducted at y_c = 0.3 to 0.8 for both the low-temperature type and the high-temperature type by increasing the section of the low-temperature air flow rate (y_c). To make a difference from the indirect heat exchange method, a heater core that does heat exchange was not used. Figures 25 and 26 are the results of experiments conducted in a direct manner in which compressed air, which has undergone temperature separation, enters the chamber directly and is discharged through an air conditioning filter. Through the experiments, the value $y_c = 0.6$, which indicates the best efficiency at the exit of the low-temperature region, led to a value of T_{co} (°C) = 21.76 °C when $P_i = 0.5$ kgf/cm^2; this changed to 9.87 °C when $P_i = 5.0$ kgf/cm^2, resulting in a maximum of 11.89 °C ΔT_{ah} (°C). Through this direct heat exchange method, the low-temperature airflow ratio (y_c) did not change between the indirect and direct heat exchange methods. As a result of an experiment that involved adding $y_c = 0.3, 0.7$, and 0.8 of T_{co} (°C) in the direct heat exchange method, it was confirmed that the efficiency decreased as the pressure increased because the flow rate was insufficient in the low-temperature region, similar to the cases of $y_c = 0.6, 0.7$. It was confirmed that T_{ho} (°C) produces a high-temperature region and has a resulting value similar to that of the indirect heat exchange method and similar to the value of the low-temperature region. The efficiency at T_{ho} (°C) was the best at $y_c = 0.6$; when $P_i = 0.5$ kgf/cm^2, it was 22.97 °C. Additionally, when $P_i = 5.0$ kgf/cm^2, it changed to 34.66 °C and showed a maximum temperature difference of 11.63 °C. Comparing the two areas, when the ratio of the discharge flow rate became high ($y_c = 0.3$ to 0.4), it showed a trend similar to that in the basic experiment. This is because the supply pressure increases as the discharge flow rate increases in the high-temperature region, thereby increasing the flow rate discharged to the high-temperature outlet stage. Therefore, heat loss in the high-temperature chamber decreased, and the amount of temperature change decreased.

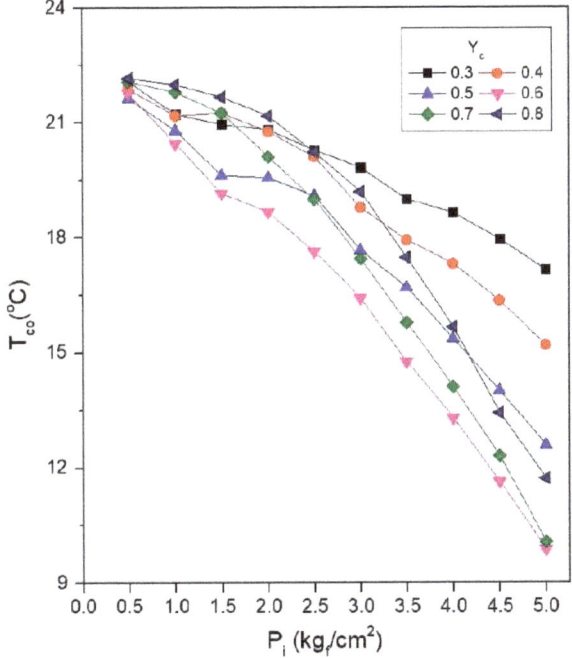

Figure 25. Direct heat exchange method of ΔT_{co} according to pressure.

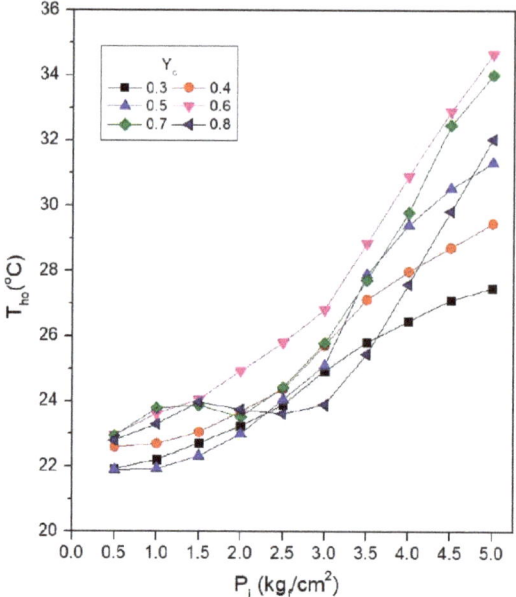

Figure 26. Direct heat exchange method of ΔT_{ho} according to pressure.

4. Conclusions

Using a vortex tube, a basic experiment was conducted to investigate the temperature separation phenomenon according to the number of nozzles of the generator according to the supply pressure (P_i) and the low-temperature air flow rate (y_c). In basic temperature difference experiments, the following conclusions were obtained using indirect and direct methods with a heater core that can be applied to a vehicle air conditioning system.

1. When the supply pressure between $P_i = 0.5$ kgf/cm^2 and $P_i = 5.0$ kgf/cm^2, the maximum high-temperature air temperature difference ($\Delta T_{h, max}$) appeared at $y_c = 0.8$, and the maximum low-temperature air temperature difference ($\Delta T_{c, max}$) appeared at $y_c = 0.5$.

2. Using 4, 5, 6, 7, and 8 nozzles in the vortex tube generator, the maximum high-temperature air temperature difference ($\Delta T_{h, max}$) and the maximum low-temperature air temperature difference ($\Delta T_{c, max}$) appeared for the maximum number of seven nozzles of the vortex tube generator.

3. The total discharge flow rate was tested by setting the supply pressure between $P_i = 0.5$ kgf/cm^2 and $P_i = 5.0$ kgf/cm^2 at intervals of $P_i = $ kgf/cm^2 at the nozzle area ratio $S_n = 0.142$ (high temperature type). The discharge flow rate tended to increase in proportion to the pressure increase; and, as the number of nozzles of the generator decreased, the nozzle inner diameter (D_n) increased, resulting in four nozzles having the highest discharge flow rate.

4. In the indirect heat exchange method, efficiency was best at $y_c = 0.8$ in the high-temperature region, and at $y_c = 0.5$ in the low-temperature region.

5. According to the results of temperature measurement in the indirect heat exchange method, T_{aoc} (°C) changed from 21.84 °C to 9.87 °C when the pressure was between $P_i = 0.5$ kgf/cm^2 and $P_i = 5.0$ kgf/cm^2 in the low-temperature region at $y_c = 0.6$ and in the high-temperature region at $y_c = 0.8$. T_{aoh} (°C) showed a maximum change of 15.79 °C, from 23.00 °C to 38.79 °C.

6. In the direct heat exchange method, the flow rate was slower than in the indirect heat exchange method, so the discharge flow rate was large; the area of the low-temperature air flow rate (y_c) was measured at $y_c = 0.3$~0.8. Excellent results were obtained in the two regions at $y_c = 0.6$. Unlike in the

indirect method, $y_c = 0.6$ was excellent at high temperature, and seems to have been affected by the discharge flow rate.

7. According to the temperature measurement results using the direct heat exchange method, when $y_c = 0.6$ and between $P_i = 0.5$ kgf/cm^2 and $P_i = 5.0$ kgf/cm^2, T_{co} (°C) changed from 21.76 °C to 9.87 °C and its maximum difference was 11.89 °C. Additionally, T_{ho} (°C) showed a maximum change of 11.63 °C from 22.97 °C to 34.66 °C.

To conclude this study, the temperature difference between the indirect heat exchange method and the direct heat exchange method showed a maximum value of about 4 °C in the high-temperature region; the temperature difference in the low-temperature region was insignificant. Although the indirect heat exchange method looks slightly better when comparing the time to reach normal temperature, the indirect heat exchange method takes about four times longer than the direct heat exchange method, and the direct heat exchange method seems to have an excellent reaction speed. This is because the use of an indirect heat exchange type heater core induces flow velocity resistance, and the heat exchange of the heater core is performed only with air. However, when applied to an air conditioning system, the indirect heat exchange method, using external air or internal air, can implement a filter to reduce foreign substances; however, in the direct method, compressed air can be filtered only through the air conditioning filter. It is thought that it may be necessary to solve this problem. Therefore, for a vehicle air conditioning system, a direct heat exchange method with low reaction speed and low flow resistance can be used and, in the case of low-temperature airflow ratio (y_c), a value of $y_c = 0.6$, with high efficiency in both the low temperature and high-temperature regions, will subsequently be used in the air conditioning system of an eco-friendly vehicle. It is judged that this system can be applied and used effectively.

Author Contributions: All authors contributed to this research in collaboration. Y.K. have an experiment, S.I. proposed the conceptualization, and J.H. provided substantial help with the paper schedule and implemented the proposed vehicle air conditioning system using vortex tube. All authors have read and agreed to the published version of the manuscript.

Funding: This research received no external funding.

Acknowledgments: This work was supported by the National Research Foundation of Korea (NRF) grant funded by the Korean government (MSIT) (no. NRF-2019R1G1A1100739).

Conflicts of Interest: The authors declare no conflict of interest.

Nomenclature

A	The rate of vortex flow coming inside [L/min]
c	Constant
D	The inner diameter of the vortex tube [mm]
d_c	The diameter of cold end orifice [mm]
D_n	The inner diameter of nozzles [mm]
L	Tube length [mm]
P_i	The pressure of inlet air [kgf/cm^2]
Q	Mean air flow rate [L/min]
S_n	Nozzle area ratio(nozzle sectional area/vortex tube sectional area) [-]
T	Static temperature [K]
ΔT_{ac}	The temperature difference between inlet air and cold outlets air for chamber [°C] ($\Delta T_{ac} = T_{aoc} - T_{ai}$)
ΔT_{ah}	The temperature difference between inlet air and hot outlets air for chamber [°C] ($\Delta T_{ah} = T_{aoh} - T_{ai}$)
T_{aoc}	The temperature of outlets cold air for chamber [°C]
T_{aoh}	The temperature of outlets hot air for chamber [°C]
T_c	The temperature of cold air [°C]
ΔT_c	The temperature difference between inlet air and cold air [°C] ($\Delta T_c = T_c - T_o$)

$\Delta T_{c,\,max}$	The maximum temperature difference between inlet air and cold air [°C]
T_{co}	The temperature of cold outlet air for after passing the fan [°C]
T_h	The temperature of hot air [°C]
ΔT_h	The temperature difference between inlet air and hot air [°C] ($\Delta T_h = T_h - T_o$)
$\Delta T_{h,\,max}$	The maximum temperature difference between inlet air and hot air [°C]
T_{ho}	The temperature of hot outlet air for after passing the fan [°C]
T_{ai}	The temperature of inlet air for chamber [°C]
$\Delta T_{w,\,max}$	The maximum temperature difference between inlet air and hot wall [°C]
y_c	Cold air mass flow ratio (cold air mass flow/inlet air mass flow)
w	Angular velocity [L/s]
r	Diameter radial direction

Subscript

c	Cold air
h	Hot air
I	Inlet air
max	Maximum
N	Nozzle and number

References

1. Jetter, J. Evaluation of Alternatives for HFC-134a Refrigerant in Motor Vehicle Air Conditioning. In Proceedings of the International Conference on Ozone Protection Technologies, Washington, DC, USA, 21–23 October 1996; pp. 845–854.
2. *Global Environmental Change Report, A Brief Analysis of the Kyoto Protocol.* December 1997, Volume IX. No. 24. Available online: https://www.oecd.org/dev/1923191.pdf (accessed on 7 October 2020).
3. Linderstrøm-Lang, C. Gas separation in the Ranque-Hilsch vortex tube. *Int. J. Heat Mass Transf.* **1964**, *7*, 1195–1206. [CrossRef]
4. Ranque, G.J. Experiments on expansion in a vortex with simultaneous exhaust of hot air and cold air. *Phys. Radium* **1933**, 112–114.
5. Hilsch, R. The Use of the Expansion of Gases in a Centrifugal Field as Cooling Process. *Rev. Sci. Instr.* **1947**, *18*, 108–113. [CrossRef] [PubMed]
6. Fulton, C.D. Ranque's Tube. *Refrig. Eng.* **1950**, *5*, 473–479.
7. Stephan, K.; Lin, S.; Durst, M.; Huang, F.; Seher, D. An investigation of energy separation in a vortex tube. *Int. J. Heat Mass Transf.* **1983**, *26*, 341–348. [CrossRef]
8. Deissler, R.; Perlmutter, M. Analysis of the flow and energy separation in a turbulent vortex. *Int. J. Heat Mass Transf.* **1960**, *1*, 173–191. [CrossRef]
9. Kassner, R.; Knoernschild, E. *Friction Laws and Energy Transter in Circular Flow*; Wright Patterson Air Force Base: Montgomery County, OH, USA, 1948; p. 259.
10. Behera, U.; Paul, P.; Kasthurirengan, S.; Karunanithi, R.; Ram, S.; Dinesh, K.; Jacob, S. CFD analysis and experimental investigations towards optimizing the parameters of Ranque–Hilsch vortex tube. *Int. J. Heat Mass Transf.* **2005**, *48*, 1961–1973. [CrossRef]
11. Gao, C.M.; Bosschaart, K.J.; Zeegers, J.C.H.; de Waele, A.T.A.M. Experimental study on a simple Ranqe-Hilsch vortex tube. *Cryogenics* **2005**, *45*, 173–183. [CrossRef]
12. Westley, R. The College of Aeronautics (Cranfield) Note, No. 30(1955-5): No. 67(1957-7). Available online: https://www.cambridge.org/core/journals/aeronautical-journal/article/research-at-the-college-of-aeronautics-cranfield/E5F9971CA8C27168C0349632BF1AA963 (accessed on 7 October 2020).
13. Hatnett, J.P.; Eckert, E.R.G. Experimental Study of the Velocity and Temperature Distribution in a High Velocity Vortex-type Flow. *Thans. ASME* **1957**, *79*, 751–758.
14. Takahama, H. Studies on Vortex Tubes (2nd Report), Reynolds Number the Effect of the Cold Air Rate and the Partial Admission of Nozzle on the Energy Separation. *Bull. JSME* **1996**, *9*, 121–130. [CrossRef]
15. Comasser, S. The Vortex tube. *J. AM. Soc. Naval Eng.* **1951**, *63*, 99–108. [CrossRef]

16. Stephan, K.; Lin, S.; Durst, M.; Huang, F.; Seher, D. A similarity relation for energy separation in a vortex tube. *Int. J. Heat Mass Transf.* **1984**, *27*, 911–920. [CrossRef]
17. Fröhlingsdorf, W.; Unger, H. Numerical investigations of the compressible flow and the energy separation in the Ranque–Hilsch vortex tube. *Int. J. Heat Mass Transf.* **1999**, *42*, 415–422. [CrossRef]
18. Eiamsa-Ard, S.; Promvonge, P. Numerical investigation of the thermal separation in a Ranque–Hilsch vortex tube. *Int. J. Heat Mass Transf.* **2007**, *50*, 821–832. [CrossRef]
19. Skye, H.M.; Nellis, G.F.; Klein, S.A. Comparison of CFD analysis to empirical vortex tube. *Int. J. Refrig.* **2006**, *29*, 71–80. [CrossRef]
20. Kaya, H.; Uluer, O.; Kocaoğlu, E.; Kirmaci, V. Experimental analysis of cooling and heating performance of serial and parallel connected counter-flow Ranquee–Hilsch vortex tube systems using carbon dioxide as a working fluid. *Int. J. Refrig.* **2019**, *106*, 297–307. [CrossRef]
21. Li, N.; Jiang, G.; Fu, L.; Tang, L.; Chen, G. Experimental study of the impacts of the cold mass fraction on internal parameters of a vortex tube. *Int. J. Refrig.* **2019**, *104*, 151–160. [CrossRef]
22. Hu, Z.; Li, R.; Yang, X.; Yang, M.; Day, R.; Wu, H.-W. Energy separation for Ranque-Hilsch vortex tube: A short review. *Therm. Sci. Eng. Prog.* **2020**, *19*, 100559. [CrossRef]
23. Rafiee, S.E.; Sadeghiazad, M.M. Experimental analysis on the impact of navigator's angle on velocimetry and thermal capability of RH-vortex tube. *Appl. Therm. Eng.* **2020**, *169*, 114907. [CrossRef]
24. Fotso, B.M.; Talawo, R.; Nguefack, M.F.; Fogue, M. Modeling and thermal analysis of a solar thermoelectric generator with vortex tube for hybrid vehicle. *Case Stud. Therm. Eng.* **2019**, *15*, 100515. [CrossRef]
25. Thakare, H.; Parekh, A.D. Experimental investigation of Ranque—Hilsch vortex tube and techno—Economical evaluation of its industrial utility. *Appl. Therm. Eng.* **2020**, *169*, 114934. [CrossRef]
26. Dutta, T.; Sinhamahapatra, K.; Bandyopadhyay, S.S. CFD Analysis of Energy Separation in Ranque-Hilsch Vortex Tube at Cryogenic Temperature. *J. Fluids* **2013**, *2013*, 1–14. [CrossRef]
27. Im, S.Y. An Experimental Study on the Geometric setup of a Low-Pressure Vortex tube. Master's Thesis, Chungnam National University, Daejeon, Korea, 2004.

© 2020 by the authors. Licensee MDPI, Basel, Switzerland. This article is an open access article distributed under the terms and conditions of the Creative Commons Attribution (CC BY) license (http://creativecommons.org/licenses/by/4.0/).

MDPI
St. Alban-Anlage 66
4052 Basel
Switzerland
Tel. +41 61 683 77 34
Fax +41 61 302 89 18
www.mdpi.com

Energies Editorial Office
E-mail: energies@mdpi.com
www.mdpi.com/journal/energies

www.ingramcontent.com/pod-product-compliance
Lightning Source LLC
LaVergne TN
LVHW070153120526
838202LV00013BA/1059